MODERN ASTRODYNAMICS

Civilised man, or so it seems to me, must feel that he belongs somewhere in space and time; that he consciously looks forward and looks back.

Kenneth Clark, Civilisation (1969)

Modern Astrodynamics

FUNDAMENTALS AND
PERTURBATION METHODS

Victor R. Bond and
Mark C. Allman

PRINCETON UNIVERSITY PRESS

Library of Congress Cataloging-in-Publication Data

Bond, Victor R., 1934–
Modern astrodynamics: fundamentals and perturbation methods /
Victor R. Bond, Mark C. Allman.
p.   cm.
Includes bibliographical references and index.
ISBN 0-691-04459-7 (cloth: alk. paper)
1. Astrodynamics.   I. Allman, Mark C., 1958– .   II. Title.
TL1050.B66   1996   95-31024
629.4′1—dc20

This book has been composed in Times Roman

Princeton University Press books are printed on
acid-free paper and meet the guidelines for permanence and
durability of the Committee on Production Guidelines for
Book Longevity of the Council on Library Resources

Printed in the United States of America
by Princeton Academic Press

10   9   8   7   6   5   4   3   2   1

CONTENTS

II   Perturbation Methods

PREFACE

This book is based upon some class notes that were developed over a period of about 12 years during which the first author (VRB) taught graduate courses in astrodynamics at the University of Houston–Clear Lake (UHCL). Two of these courses were (and are) called "Fundamentals of Astrodynamics" and "Perturbation Methods." For the first few years of this period, VRB used textbooks written by other authors. Gradually the two courses evolved to take on the flavor of the astrodynamics used in the geographical area surrounding UHCL, the Clear Lake area of Houston where NASA's Johnson Space Center (JSC) and supporting aerospace contractors are located.

In 1991 we decided to write this book, and we were faced with a dilemma: How do we develop a book which is fundamentally and scientifically sound, deeply based in the rich tradition of its parent, Celestial Mechanics, and at the same time tailor it to the local situation in which pragmatism and robustness of method are very important? We hope that this book solves the dilemma.

We developed this book directly from the three laws of motion of Isaac Newton and his law of gravitation. We have taken great care not to assume that we already know the results of our development and not to rely upon intuition. On the other hand, we have recognized that the results must "work" when applied. To this extent we present our important results in the form of algorithms, going to great length to describe their limitations and where they fail when programmed.

Note from VRB

This book has been influenced strongly by the new ideas in analytical and numerical methods that began entering the field of celestial mechanics after about 1962. In particular we refer to the period from about 1970 to about 1982 when several visiting scientists and mathematicians were located at JSC (formerly the Manned Spacecraft Center) working under various funding arrangements and for various intervals of time in areas such as regularization, the perturbed two-body problem with redundant variables, Hamiltonian mechanics in extended phase space, the restricted three-body problem, stability, and numerical methods. This work was conducted under Jack Funk, Victor Bond, and Don Jezewski of the Mission Planning and Analysis Division (MPAD) of JSC. The scientists and mathematicians in this group (and their institutional affiliations) were as follows: Victor Szebehely, University of Texas; Dale Bettis, University of Texas; Roger Broucke, University of Texas; Otis Graf, University of Texas; Paul Nacozy, University of Texas; Kathleen Horn, University of Texas; Alan Mueller, University of Texas; John Donaldson, University of Tasmania; Gerhard Scheifele, Swiss Federal Institute (Zurich); Guy Janin, European Space Agency (Darmstadt); Gordon Johnson, University of Houston; Victor Bogdan, Catholic University; Euell Kennedy, California Polytechnic State University; Don Mittleman, Oberlin College; Bruce Feiring, University of Minnesota; Robert Sartain, Houston Baptist University; Murray Silver, Bowdoin College; A. A. Jackson, Saint Thomas University.

I also wish to thank my wife Peggy for her support during the preparation of this book. She is beautiful, a first-class mother and grandmother, and a talented artist. Without her patience, understanding, and advice, this book might not have been written.

Note from MCA

The topics presented in this book are (to a large part) used in the "real world," and as such are presented from an "applications" point of view. However, these topics have a rich history (which we touch on from time to time) in classical mechanics. Digging into the question, "How did we know to do something, try a particular solution, or use a certain method?" involves digging into the physics on which astrodynamics is based. For example, in the two-body problem no external forces act on the system. This immediately tells us that the energy and angular momentum are conserved (constant). Analysis of systems involves understanding concepts such as, for example, symmetry, invariance, and covariance, and theorems developed in advanced mechanics allow us, for example, to find conservation laws for dynamical systems (Noether's theorem). If it appears that we "pull an idea or relationship out of the air," what is actually happening is recognizing a property (e.g., symmetry) of the system. We strongly encourage students to pursue the study of Hamiltonian mechanics, dynamical systems, and group theory to build on what they learn here.

This book was first developed using T_EX, then later LAT_EX, on a NeXT workstation. NeXTStep provides an excellent environment for putting the book together (and taking it apart again), allowing us to concentrate on content and not on mechanics.

For her patience, understanding, assistance with everything resembling punctuation rules ("...end sentences in periods? *Really?*"), periodic advice on manuscript production ("Did you put paper in the printer?"), and unending encouragement, a special thanks is in order to my wife Janice. And of course I cannot fail to credit the (indispensable) aid of my two cats—without their assistance the work would've required (at most) half the time.

SPECIAL THANKS go to Trevor Lipscombe and Alice Calaprice at Princeton University Press for their infinite patience in answering our endless barrage of e-mail and their professional handling of two rather anxious authors. We are also grateful for the support of the faculty and administration of the University of Houston–Clear Lake; especially to Robert Hopkins, Theron Garcia, James Lester, George Blanford, Ronald Mills, Carroll Lassiter, Edward Dickerson, Terry Feagin, Edward Hugetz, and Norm Richert. Finally, we would like to thank all those students who used preliminary versions of the text and pointed out the many errors and omissions.

April 1995

I

FUNDAMENTALS

1

BACKGROUND

1.1 Introduction

The primary focus of this book is to introduce the two-body problem by developing its differential equations from Newton's laws of motion and Newton's universal law of gravitation. The geometry of the problem will develop from the integrals of the motion. In particular we will not use the famous empirical laws of Kepler until we have developed the solution of the two-body problem in terms of the integrals of the motion to a point where we can verify these laws.

Few numerical problems will appear in this book. It is our feeling that the rigid development of the differential equations of motion from physical laws and the development of the solution to these equations are more important than numerical problems. However, the algorithms necessary for computing the solutions of the two-body problem are presented throughout this book. In fact, the algorithms will be numbered consecutively and are included in the table of contents. These solutions and their algorithms will be of both the initial-value type (that is, given initial position and velocity at initial time, find the position and velocity at some other time) and the two-point boundary value type (Lambert's problem, for example).

1.2 Notation and Units

We choose a notation that we believe is as simple as possible. The normal rule is that a vector will be in bold-face type; for example, the position vector $\mathbf{r}$. The magnitude of this vector is

$$r = \|\mathbf{r}\|$$
$$= \sqrt{\mathbf{r} \cdot \mathbf{r}},$$

and the unit vector in the direction of **r** is

$$\hat{\mathbf{r}} = \frac{1}{r}\mathbf{r}.$$

There are exceptions to this rule. For example, the velocity vector is

$$\mathbf{v} = \dot{\mathbf{r}},$$

but the magnitude of this vector is

$$v = \|\dot{\mathbf{r}}\|.$$

Note, however, that, in general, $\dot{r} \neq \|\dot{\mathbf{r}}\|$. The unit vector corresponding to **v** is

$$\hat{\mathbf{v}} = \frac{1}{\|\dot{\mathbf{r}}\|}\dot{\mathbf{r}}.$$

The term "velocity vector" is somewhat redundant since "velocity" alone implies a direction. In general, derivatives of vectors must be renamed before the normal rules would apply.

Another exception is the unit vector $\hat{\boldsymbol{\phi}}$, which is in the direction of the local horizontal. There is no vector $\vec{\boldsymbol{\phi}}$ nor is there a magnitude ϕ related to $\hat{\boldsymbol{\phi}}$. In fact, the symbol ϕ, used alone, will be used exclusively as an angle.

Throughout this book we will quite often require that partial derivatives involving vectors be taken. For example, the partial derivatives of a vector with respect to a scalar is a vector. Let

$$\mathbf{v} = \mathbf{v}(a);$$

then

$$\frac{\partial \mathbf{v}}{\partial a} = \mathbf{w}$$

is another vector.

The partial derivatives of a scalar with respect to a vector is a vector. That is, if

$$a = a(\mathbf{v}),$$

then

$$\frac{\partial a}{\partial \mathbf{v}} = \mathbf{x}$$

is another vector.

The partial derivatives of a vector with respect to a vector is a matrix. Let

$$\mathbf{v} = \mathbf{v}(\mathbf{u}),$$

which is a vector $\mathbf{v}$ with n components which is a function of a vector $\mathbf{u}$ with m components. Then

$$\frac{\partial \mathbf{v}}{\partial \mathbf{u}} = \mathbf{M} = \begin{pmatrix} \partial v_1/\partial u_1 & \partial v_1/\partial u_2 & \cdots & \partial v_1/\partial u_m \\ \partial v_2/\partial u_1 & \partial v_2/\partial u_2 & \cdots & \partial v_2/\partial u_m \\ \vdots & \vdots & \ddots & \vdots \\ \partial v_n/\partial u_1 & \partial v_n/\partial u_2 & \cdots & \partial v_n/\partial u_m \end{pmatrix}.$$

Occasionally vectors will be considered to be either column or row vectors. For example,

$$\mathbf{v} = \begin{pmatrix} v_1 \\ v_2 \\ \vdots \\ v_n \end{pmatrix}$$

is a column vector, but

$$\mathbf{v}^T = (v_1, v_2, \ldots, v_n)$$

is a row vector. Using this approach we could also write the partial derivatives of the scalar

$$a = a(\mathbf{v})$$

as

$$\frac{\partial a}{\partial \mathbf{v}} = \mathbf{x}^T.$$

Wherever the meaning is clear we will not distinguish between column and row vectors. Obviously, whenever vector matrix operations are being discussed, the concept of column and row vectors is important. For example, consider the transformation

$$\mathbf{x} = N\mathbf{y},$$

where $\mathbf{x}$ and $\mathbf{y}$ are column vectors and N is a matrix. Then the expression

$$\mathbf{x}^T = \mathbf{y}^T N^T$$

is clearly valid and expresses exactly the same information as the previous relation except that $\mathbf{x}^T$ and $\mathbf{y}^T$ are row vectors.

Also, in some instances the dot product and transpose notations are interchangeable:

$$\mathbf{v} \cdot \mathbf{w} = \mathbf{v}^T \mathbf{w}.$$

For the case where two vectors

$$\mathbf{a} = \hat{\mathbf{i}}\,a_1 + \hat{\mathbf{j}}\,a_2 + \hat{\mathbf{k}}\,a_3$$

and

$$\mathbf{b} = \hat{\mathbf{i}}\,b_1 + \hat{\mathbf{j}}\,b_2 + \hat{\mathbf{k}}\,b_3$$

are defined in a cartesian coordinate system $(\hat{\mathbf{i}}, \hat{\mathbf{j}}, \hat{\mathbf{k}})$, their cross product can be represented as a vector-matrix operation. That is,

$$\mathbf{a} \times \mathbf{b} = A\,\mathbf{b},$$

where A is the matrix

$$A = \begin{pmatrix} 0 & -a_3 & a_2 \\ a_3 & 0 & -a_1 \\ -a_2 & a_1 & 0 \end{pmatrix}.$$

The dot or scalar product of two vectors $\mathbf{a}$ and $\mathbf{b}$ gives the scalar

$$\mathbf{a} \cdot \mathbf{b} = ab\cos\theta,$$

where $a = \|\mathbf{a}\|$, $b = \|\mathbf{b}\|$ and θ is the angle between $\mathbf{a}$ and $\mathbf{b}$. The cross or vector product gives another vector,

$$\mathbf{a} \times \mathbf{b} = ab\sin\theta\,\hat{\mathbf{l}},$$

where a, b, and θ have the same definition as the dot product and $\hat{\mathbf{l}}$ is a unit vector normal to the plane containing $\mathbf{a}$ and $\mathbf{b}$. That is,

$$\hat{\mathbf{l}} = \frac{\mathbf{a} \times \mathbf{b}}{\|\mathbf{a} \times \mathbf{b}\|}.$$

The cross product of two vectors can also be obtained by the "determinant" method,

$$\mathbf{a} \times \mathbf{b} = \begin{vmatrix} \hat{\mathbf{i}} & \hat{\mathbf{j}} & \hat{\mathbf{k}} \\ a_1 & a_2 & a_3 \\ b_1 & b_2 & b_3 \end{vmatrix}$$

$$= \hat{\mathbf{i}}(a_2 b_3 - a_3 b_2) + \hat{\mathbf{j}}(a_3 b_1 - a_1 b_3) + \hat{\mathbf{k}}(a_1 b_2 - a_2 b_1).$$

Several vector relationships will be used repeatedly throughout this text. For example, the cross product of two vectors is not commutative,

$$\mathbf{a} \times \mathbf{b} = -\mathbf{b} \times \mathbf{a}.$$

There are two forms for the *triple vector product*,

$$(\mathbf{a} \times \mathbf{b}) \times \mathbf{c} = (\mathbf{a} \cdot \mathbf{c})\mathbf{b} - (\mathbf{b} \cdot \mathbf{c})\mathbf{a}$$

and

$$\mathbf{a} \times (\mathbf{b} \times \mathbf{c}) = (\mathbf{a} \cdot \mathbf{c})\mathbf{b} - (\mathbf{a} \cdot \mathbf{b})\mathbf{c}$$

Clearly, in the triple vector product placement of the parentheses is significant. The *triple scalar product* is given by

$$\mathbf{a} \cdot \mathbf{b} \times \mathbf{c} = \mathbf{a} \times \mathbf{b} \cdot \mathbf{c} = \mathbf{c} \cdot \mathbf{a} \times \mathbf{b} = \mathbf{c} \times \mathbf{a} \cdot \mathbf{b}.$$

Note that the dot and cross operators can be interchanged as long as the order of the vectors is maintained. Also note that the triple scalar product is zero if any two vectors are parallel, since the cross product term is zero for parallel vectors. This is important since then *any* triple scalar product where two of the vectors are the same is zero:

$$\mathbf{a} \cdot \mathbf{b} \times \mathbf{a} = 0.$$

The triple scalar product can also be found from the determinant,

$$\mathbf{a} \cdot \mathbf{b} \times \mathbf{c} = \begin{vmatrix} a_1 & a_2 & a_3 \\ b_1 & b_2 & b_3 \\ c_1 & c_2 & c_3 \end{vmatrix}.$$

Another useful identity is the dot product of two cross products,

$$(\mathbf{a} \times \mathbf{b}) \cdot (\mathbf{c} \times \mathbf{d}) = (\mathbf{b} \cdot \mathbf{d})(\mathbf{c} \cdot \mathbf{a}) - (\mathbf{b} \cdot \mathbf{c})(\mathbf{d} \cdot \mathbf{a}).$$

Since only a few numerical problems occur in this book, systems of units should not be a problem. However, except for one example, the units used will be SI units, also known as MKSA units, as defined by the U.S. National Bureau of Standards [41]. One of the most interesting aspects in this NBS report is that conversions from British units to SI units do not appear. There is a subtle message to all of us in this omission. The International Astronomical Union (IAU) approved a list of constants [45][2] which are reproduced in tables 1.1, 1.2, 1.3, and 1.4. The astronomical constants are classified [2] into three types. The constants in the first class are called the *defining constants* and are given in table 1.1. As their name suggests, their values are static and no change in their values is anticipated. For example, the Gaussian gravitational constant (k) has the same value as that derived by Gauss in 1809.

Table 1.1 Defining constants.

Description	Symbol	Value	Units
Gaussian gravitational constant	k	0.01720209895	1/day
Speed of light	c	299792.458	km/sec

Table 1.2 Primary constants.

Description	Sym.	Value	Units
Light time for unit distance	τ_A	499.004782	sec
Equatorial radius for the Earth	ae	6378.140	km
Dynamical form-factor for Earth	J_2	0.00108263	
Geocentric gravitational constant	GE	398600.5	km^3/sec^2
Constant of gravitation	G	6.672×10^{-20}	$km^3/(kg\ sec^2)$
(mass of Moon)/(mass of Earth)	μ	0.01230002	
General precession in longitude, per Julian century, at J2000	p	5029.0966	arc-sec.
Obliquity of ecliptic at J2000	ϵ	23.439291111	deg.

Table 1.3 Derived constants.

Description	Sym.	Value	Units
Constant of nutation at J2000	N	9.2025	arc-sec.
Unit distance	A	149597870	km
Solar parallax	π_{sun}	8.794148	arc-sec.
Constant of aberration at J2000	κ	20.49552	arc-sec.
Flattening factor for Earth	f	1/298.257	
Heliocentric Grav. constant	GS	$1.32712438 \times 10^{11}$	km^3/sec^2
(mass of Sun)/(mass of Earth)	S/E	332946.0	
(mass of Sun)/(mass of Earth + Moon)		328900.5	
Mass of Sun	S	1.9891×10^{30}	kg

The constants in the second class are called *primary constants*. The values of these constants are determined from observations, and are listed in table 1.2. Note that the constant of gravitation (G) is known only to four significant digits. Its value has changed only in the fourth digit since it was first measured by Cavendish in 1798. Because of this imprecision, the constant of gravitation should be used with care in calculations.

The constants in the third class are called *derived constants*. These are determined from the defining constants and the primary constants. There are many constants in this class; nine of the most important are given in table 1.3.

Table 1.4   Mass ratios of sun to planets.

Mercury	Venus	Mars	Jupiter
6023600	408523.5	3098710	1047.355

Saturn	Uranus	Neptune	Pluto
3498.5	22869	19314	3000000

Some other derived constants, the ratio of the mass of the Sun to the mass of each planet, are given in table 1.4.

Note that we can use the heliocentric gravitational constant (GS) from table 1.3 and a mass ratio from table 1.4 to compute the gravitational constant for a planet. For example, let J be the mass of Jupiter and S be the mass of the Sun. From tables 1.3 and 1.4 we then calculate Jupiter's gravitational constant,

$$\frac{S}{J} = 1047.355$$

$$\frac{GS}{GJ} = 1047.355$$

$$GJ = \frac{GS}{1047.355}$$

$$= \frac{1.32712438 \times 10^{11} \text{ km}^3/\text{sec}^2}{1047.355}$$

$$= 1.2671203 \times 10^8 \text{ km}^3/\text{sec}^2.$$

1.3 Time

In this text we adopt the definition of time given by Lindsay and Margenau [50]: "Abstract time as it appears in the equations of physics is merely a parameter which serves as a useful independent variable and whose range of variation is the real number continuum." This definition is very close to that given in the Explanatory Supplement to the Ephemeris [31]: "Ephemeris Time is a uniform measure of time depending for its determination on the laws of dynamics. It is the independent variable in the gravitational theories of the Sun, Moon and planets, and the argument for the fundamental ephemerides in the Ephemeris."

Time as defined in [31] is called Ephemeris Time (ET), whose introduction fired a shot across the bow of a battleship manned by able-minded Einsteinians and other post-Newtonian theorists. Neither of these definitions faces the problem of relating the solutions of the equations of physics at some value of time with the physical reality of how time is measured. The solution of this problem is well beyond the scope of this book. The "time" which we shall use is equivalent to Terrestrial Dynamical Time (TDT), which was introduced in 1977 (Astronomical Almanac [2]). For all practical purposes TDT is the same as the abstract time of Lindsay and Margenau in 1957 and the ET of the Explanatory Supplement of 1961.

In addition to TDT there are other timescales of which the student should be aware, even though they are not emphasized in this book. The first three of the other scales given below are defined in the 1985 Astronomical Almanac [2]. The fourth scale is inferred from a Time Service Announcement [74].

- International Atomic Time (TAI) is related to TDT by

$$\text{TDT} = \text{TAI} + 32.184 \text{ sec.}$$

- Coordinated Universal Time (UTC), which is the basis for all civil and legal timescales, is related to TAI by

$$\text{TAI} = \text{UTC} + \Delta\text{AT},$$

where ΔAT is the accumulated number of "leap seconds" since 1971. ΔAT is subject to change by an exact second twice a year. Note that TDT and TAI are uniform timescales, but UTC by its definition is discontinuous.

- Universal Time (UT1) is defined by the Greenwich Mean Sidereal Time (GMST) at midnight, that is,

$$\text{GMST (at UT1} = 0) = 24110.54841 + (8640184.812866)Tu$$
$$+ (0.093104)Tu^2 - (6.2 \times 10^{-6})Tu^3,$$

where the units are seconds and Tu is defined by

$$Tu = \frac{(\text{Julian Date @ Midnight}) - 2451545}{36525},$$

which is the fraction of the Julian century (36525 days of Universal Time) from the Standard Epoch J2000 (Julian date 2451545.0 UT1) to midnight expressed as a Julian date. UT1 is related to UTC by

$$\text{UT1} = \text{UTC} + \text{DUT1}.$$

The quantity DUT1 is determined by observation but is constrained such that the difference

$$|\text{UT1} - \text{UTC}| < 0.9 \text{ sec.}$$

Whenever a violation of this constraint is anticipated, the number of leap seconds, ΔAT, is adjusted by exactly one second. The quantities ΔAT and DUT1 are published weekly by the United States Naval Observatory [40].

- The Global Positioning System timescale (GPST) is a uniform timescale related to TAI by

$$TAI = GPST + 19 \text{ sec.}$$

Note that we have introduced four equations relating the five timescales (TDT, TAI, UTC, UT1, and GPST). If any one of these timescales is chosen to be the independent variable of the equation of motion, then the other four timescales can be determined if the quantities ΔAT and UT1 are known.

In the above relations we mean the unit of "seconds" to be the same as the international second defined in the NBS publication [41], which is based on radiation levels of the Cesium-133 atom. We also note that the Julian date is defined as the interval in days and fractions of a day since 1 January 4713 B.C., Greenwich noon. In precise work, the timescale (TDT or UT1) should be specified.

Finally, we present an expression for the rotational rate of the Earth as given by the Explanatory Supplement to the Astronomical Almanac [59]:

The duration (Λ) of the day in Universal Time is

$$\Lambda = 86400 - \frac{\psi_2 - \psi_1}{n},$$

where ψ_1 and ψ_2 are the values of UT1 $-$ TAI in SI seconds at n-day intervals. The angular velocity (i.e., rotation rate) of the Earth is given by

$$\omega = \frac{86400}{\Lambda}(72.921151467 \times 10^{-6} \text{ radians/sec}).$$

This expression accounts for the irregular rotation of the Earth. From the above definitions we see that

$$UT1 - TAI = (UT1 - UTC) - \Delta AT,$$

where the difference UT1 $-$ UTC is always less than 0.9 sec in magnitude and ΔAT is the accumulated number of leap seconds. Note that the factor $86400/\Lambda$ is very nearly one.

2

THE TWO-BODY PROBLEM

2.1 Newton's Laws

In this section we derive the differential equations of motion of the two-body problem. This derivation starts with the physics of the problem that involves the application of Newton's laws of motion and Newton's universal law of gravitation. Newton's laws of motion [50][27] can be stated as follows:

1. A particle of mass remains at rest or moves in a straight line at a constant speed until acted upon by a force.
2. The force acting on a particle of mass is proportional to the total time derivative of the product of the mass and the velocity of the particle.
3. When two particles of mass A and B interact, the force exerted by mass A on mass B is equal in magnitude and opposite in direction to the force exerted by mass B on mass A.

Newton's universal law of gravitation describes the force of attraction between two particles of mass as a force directly proportional to the product of their masses and inversely proportional to the square of the distance of separation.

With these laws in hand we define an inertial frame of reference, using the most common definition, to be one in which Newton's laws of motion are valid.

Since the approach taken is to derive the fundamental integrals, we next give the classical definition of an integral of the motion and briefly discuss the concept of a completely solvable or integrable system. We then develop the fundamental integrals and relationships among them. Throughout we minimize reliance on geometry, stressing a somewhat rigid mathematical development.

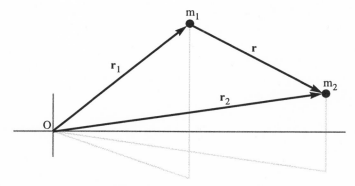

Figure 2.1   Two points of mass relative to fixed origin O.

2.2 Physics of the Two-Body Problem

We consider two points of mass, m_1 and m_2, located at positions $\mathbf{r}_1$ and $\mathbf{r}_2$ with respect to a fixed origin O. By a fixed origin we mean a point that is fixed to an inertial frame, which we defined in §2.1. We assume that these two masses interact by a force that is a function only of the relative distance between the two masses and that is directed along the position of m_2 with respect to m_1 (fig. 2.1). This force can be expressed as

$$\mathbf{F} = F\,\hat{\mathbf{r}}, \tag{2.1}$$

where

$$\mathbf{r} = \mathbf{r}_2 - \mathbf{r}_1 \tag{2.2}$$

and the unit vector is

$$\hat{\mathbf{r}} = \frac{1}{r}\,\mathbf{r}. \tag{2.3}$$

Applying Newton's second law of motion to m_2 gives

$$m_2\ddot{\mathbf{r}}_2 = -\mathbf{F}, \tag{2.4}$$

and Newton's second and third laws of motion to m_1 yield

$$m_1\ddot{\mathbf{r}}_1 = \mathbf{F}. \tag{2.5}$$

Now obtain the differential equation of motion of m_2 relative to m_1 by subtracting (2.5) from (2.4) after dividing by their masses. The result is

$$\ddot{\mathbf{r}}_2 - \ddot{\mathbf{r}}_1 = -\frac{1}{m_2}\mathbf{F} - \frac{1}{m_1}\mathbf{F}.$$

Using equations (2.1) and (2.2) this becomes

$$\ddot{\mathbf{r}} = -\left(\frac{1}{m_2} + \frac{1}{m_1}\right) F\,\hat{\mathbf{r}}. \qquad (2.6)$$

In classical mechanics the reciprocal of the factor

$$\frac{1}{m_2} + \frac{1}{m_1} = \frac{m_1 + m_2}{m_1 m_2}$$

is called the *reduced mass*. We will not bother giving it a symbol here.

Now we invoke Newton again. This time we specify that the magnitude of the force $\mathbf{F}$ is

$$\|\mathbf{F}\| = F = G\frac{m_1\,m_2}{r^2}. \qquad (2.7)$$

That is, the force is that of Newton's universal law of gravitation. G is a proportionality factor called the Universal Gravitational Constant. Using equation (2.7) in equation (2.6) we obtain

$$\ddot{\mathbf{r}} = -\frac{m_1 + m_2}{m_1\,m_2}\,G\frac{m_1 m_2}{r^2}\,\hat{\mathbf{r}}.$$

Note the obvious cancelation and use equation (2.3) to obtain

$$\ddot{\mathbf{r}} + \frac{\mu}{r^3}\,\mathbf{r} = 0, \qquad (2.8)$$

where we have defined the factor

$$\mu = G(m_1 + m_2).$$

Equation (2.8) is the differential equation of motion (DEOM) for the two-body problem.

Comments

- Note that equation (2.8) is unchanged if we replace $\mathbf{r}$ with $-\mathbf{r}$. Thus equation (2.8) gives motion of m_2 relative to m_1 or the motion of m_1 relative to m_2. Also observe that if we replace t with $-t$ equation (2.8) remains unchanged.

- Note that equation (2.8) gives us the motion of a unit mass (m_u) relative to the mass ($m_1 + m_2$) located at a fixed origin, that is,

$$m_u \ddot{\mathbf{r}} + m_u \frac{\mu}{r^3} \mathbf{r} = 0.$$

This equation has the form of Newton's second law. Therefore, by our definition of a fixed origin, each mass moves as if it were a unit mass attracted by mass ($m_1 + m_2$) located at a distance r.

- Observe that the gravitational force is derivable from a potential. That is, equation (2.8) can be written

$$\ddot{\mathbf{r}} = -\frac{\partial V}{\partial \mathbf{r}},$$

where

$$V = -\frac{\mu}{r}.$$

- The gravitational potential is defined such that it is zero at an infinite distance from the mass at the "fixed" center. That is,

$$\lim_{r \to \infty} V = \lim_{r \to \infty} \left(-\frac{\mu}{r} \right) = 0.$$

The gravitational potential is also defined to be negative such that its value becomes smaller as r decreases. Thus the gravitational potential is always at a maximum when its value is zero.

- Note that at no point in the derivation was the concept of center of mass used.

2.2.1 Uniform Spherical Mass Potential

In the derivation above for the two-body equation of motion, we considered m_1 and m_2 to be point masses. We will now show that equation (2.8) is valid for the case where m_1 is a finite solid homogeneous sphere of radius R and m_2 is a point mass. As noted above we can reverse the roles of m_1 and m_2. What we discuss here is a general characteristic of central force potentials. Those students familiar with undergraduate electromagnetic theory will recognize the following from electric field and electrostatic potential calculations for charged spheres and shells [1].

We first derive the potential for a mass shell of radius r, thickness dr, with uniform density ρ. We will then sum all shells from $r = 0$ to $r = R$ to arrive at the total potential. Taking the negative gradient of this potential gives us the gravitational force. The spherical symmetry of the problem naturally suggests we use spherical coordinates (a, θ, ψ), which we diagram in figure 2.2.

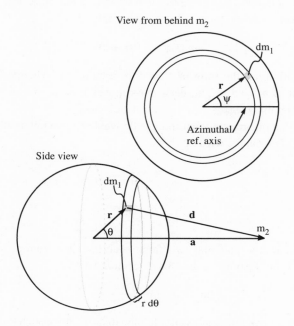

Figure 2.2 Geometry for the spherical shell potential problem.

We first work out the equation for the potential V_s at m_2 due to a spherical shell of radius r and thickness dr. We start with the definition for the infinitesimal potential,

$$dV = -\frac{G dm_1}{\|\mathbf{d}\|}.$$

The mass dm_1 is given by $dm_1 = \rho dv$, where ρ is the density (assumed constant for this problem) and dv is a unit volume. In spherical coordinates the unit volume is

$$dv = r^2 \sin\theta dr d\theta d\psi.$$

Note that the total mass m_1 is given by

$$m_1 = \int_0^R \int_0^\pi \int_0^{2\pi} \rho r^2 \sin\theta d\psi d\theta dr$$

$$= \left(\frac{4}{3}\right)\pi R^3 \rho,$$

which is just the density multiplied by the volume of the spherical mass.

We require an expression for $\|\mathbf{d}\|$. From figure 2.2 we see that $\mathbf{d} = \mathbf{a} - \mathbf{r}$. The magnitude (norm) of $\mathbf{d}$ is then

$$\begin{aligned}
\|\mathbf{d}\| &= \sqrt{\mathbf{d} \cdot \mathbf{d}} \\
&= \sqrt{(\mathbf{a} - \mathbf{r}) \cdot (\mathbf{a} - \mathbf{r})} \\
&= \sqrt{a^2 + r^2 - 2ar \cos \theta}.
\end{aligned}$$

Putting all this together we have the potential at m_2 for a shell with radius r and thickness dr given by

$$\begin{aligned}
V_s &= -\int_0^\pi \int_0^{2\pi} G\rho \frac{1}{\sqrt{a^2 + r^2 - 2ar \cos \theta}} r^2 \sin \theta \, d\psi \, d\theta \, dr \\
&= -2\pi G\rho r^2 dr \int_0^\pi \frac{\sin \theta}{\sqrt{a^2 + r^2 - 2ar \cos \theta}} d\theta.
\end{aligned}$$

This integral is actually quite simple. Set

$$\begin{aligned}
x &= \sqrt{a^2 + r^2 - 2ar \cos \theta} \\
dx &= \frac{ra \sin \theta}{\sqrt{a^2 + r^2 - 2ar \cos \theta}} d\theta \\
\frac{dx}{ra} &= \frac{\sin \theta}{\sqrt{a^2 + r^2 - 2ar \cos \theta}} d\theta.
\end{aligned}$$

The limits of integration change to

$$\begin{aligned}
\theta = 0 &: x = \sqrt{a^2 + r^2 - 2ar \cos 0} = a - r \\
\theta = \pi &: x = \sqrt{a^2 + r^2 - 2ar \cos \pi} = a + r
\end{aligned}$$

and the integral becomes

$$\begin{aligned}
V_s &= -2\pi G\rho r^2 dr \frac{1}{ra} \int_{(a-r)}^{(a+r)} dx \\
&= -\frac{2\pi G\rho r \, dr}{a} x \Big|_{a-r}^{a+r} \\
&= -\frac{2\pi G\rho r \, dr}{a} (a + r - a + r) \\
&= -\frac{4\pi G\rho r^2 dr}{a}.
\end{aligned}$$

An important point to note is that when evaluating the limits of the integration the value $a - r$ changes to $r - a$ for points interior to the shell, since the value is the length (norm) of a vector and therefore must be real and non-negative. Using these limits in the integration yields for an *interior* point,

$$V_s = -4\pi G\rho r\,dr = constant.$$

We now integrate the expression for V_s (*exterior* point) to give an equation for the potential for a sphere,

$$V = -\int_0^R \frac{4\pi G\rho}{a} r^2 dr$$
$$= -\frac{4}{3}\pi R^3 \frac{G\rho}{a}.$$

Recall the equation for the total mass m_1: $\frac{4}{3}\pi R^3 \rho$. We see that using this reduces the potential equation to

$$V = -\frac{Gm_1}{a},$$

which is the equation for a point mass m_1 a distance a from m_2. We can therefore treat a homogeneous mass of finite radius as a point mass, which is what we wanted to show.

We take the negative gradient of the potential to find the force on m_2 due to the gravitational potential of m_1. In spherical coordinates the gradient operator is given by

$$\nabla = \frac{\partial}{\partial a}\hat{\mathbf{a}} + \frac{1}{a\sin\theta}\frac{\partial}{\partial\psi}\hat{\boldsymbol{\psi}} + \frac{1}{a}\frac{\partial}{\partial\theta}\theta,$$

where $\hat{\mathbf{a}}$ is the radial unit vector. We see that the only non-zero term is the radial term (there is no angular dependence in the potential):

$$\mathbf{F} = -\nabla V$$
$$= -\frac{\partial V}{\partial a}\hat{\mathbf{a}}$$
$$= -\frac{Gm_1}{a^2}\hat{\mathbf{a}}.$$

The $\hat{\mathbf{a}}$ vector is directed from m_1 to m_2. The equation for $\mathbf{F}$ is the force experienced by m_2 due to m_1 and directed along the $-\hat{\mathbf{a}}$ axis. Finally, note that since the potential inside the mass shell is constant the force is 0 —a particle placed inside a mass shell experiences no gravitational force due to the shell.

2.3 Definition of an Integral

We are given (Szebehely [69]) the system of second-order differential equations having n degrees of freedom, or dimensions,

$$\ddot{\mathbf{x}} = \mathbf{F}(\mathbf{x}, \dot{\mathbf{x}}, t), \tag{2.9}$$

where the dependent variable array is

$$\mathbf{x}^T = (x_1, x_2, x_3, \ldots, x_n)$$

and the right-hand side is the array

$$\mathbf{F}^T = (f_1, f_2, f_3, \ldots, f_n).$$

Any function $G(\mathbf{x}, \dot{\mathbf{x}}, t)$ such that

$$G(\mathbf{x}, \dot{\mathbf{x}}, t) = constant, \tag{2.10}$$

where $\mathbf{x} = \mathbf{x}(t)$ is a solution of (2.9), is called an *integral of the motion* of the system (2.9). Two examples of integrals of the motion for the two-body problem are the energy and the angular momentum. These examples will be developed with others in §2.4. If $2n$ integrals exist for (2.9), then the system has a solution and is said to be completely integrable, or solvable.

Applying this to the two-body problem we note that equation (2.8) has the form of equation (2.9) if we let

$$\mathbf{x}^T = (x_1, x_2, x_3) \qquad \text{and} \qquad \mathbf{F} = -\frac{\mu}{\|\mathbf{x}\|^3} \mathbf{x}, \tag{2.11}$$

limit the size of the dependent variable array to $n = 3$, and identify $\mathbf{x}$ to be the position vector $\mathbf{r}$. We then recover equation (2.8),

$$\ddot{\mathbf{r}} = -\frac{\mu}{r^3}\mathbf{r},$$

which needs $2 \cdot 3 = 6$ integrals for a complete (or analytic) solution.

2.4 Integrals of the Two-Body Problem

At this point it is convenient to introduce an inertial frame of reference with a fixed origin located at one of the two masses, m_1 or m_2. We will locate the fixed origin at m_1 and consider the motion of m_2 with respect to m_1, although

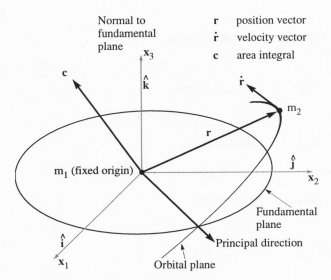

Figure 2.3 Two-body orbital motion showing the orbital plane and the area integral (**c**) in an inertial, cartesian frame.

considering the first of the comments made in §2.2 it is permissible to set up the system in the other order.

The two-body problem, equation (2.8), was derived without specifying a coordinate system. In this regard the only requirement was that the position **r** had to be measured from a fixed origin located at either m_1 or m_2. The position **r**, and for that matter equation (2.8), can be expressed in many different coordinate systems. In figure 2.3 we show an inertial cartesian frame in which the position **r** is expressed as

$$\mathbf{r} = \hat{\mathbf{i}}\,x_1 + \hat{\mathbf{j}}\,x_2 + \hat{\mathbf{k}}\,x_3,$$

where $\hat{\mathbf{i}}, \hat{\mathbf{j}}, \hat{\mathbf{k}}$ are unit vectors associated with the cartesian axes, respectively.

Inertial frames in astronomy, celestial mechanics, and related disciplines are defined by specifying a fixed origin, a fundamental plane, and a principal direction that is in the fundamental plane. In figure 2.3, m_1 is at the fixed origin, the axis x_3 (unit vector $\hat{\mathbf{k}}$) is normal to the fundamental plane, and the axis x_1 (unit vector $\hat{\mathbf{i}}$) is the fundamental direction. The third axis x_2 is formed by

$$\hat{\mathbf{j}} = \hat{\mathbf{k}} \times \hat{\mathbf{i}}.$$

The integrals of the motion for the two-body problem are developed from equation (2.8), which we manipulate using vector algebra to obtain the constants of the motion (Pollard [58]). The strategy in this development is to obtain differential equations that are total derivatives. The integration of such equations is trivial. This strategy works nicely for all but one of the six integrals of motion.

2.4.1 Area Integral c

Rewriting equation (2.8) for reference,

$$\ddot{\mathbf{r}} + \frac{\mu}{r^3}\,\mathbf{r} = 0, \tag{2.12}$$

we take the cross product of $\mathbf{r}$ with equation (2.12) to obtain

$$\mathbf{r} \times \ddot{\mathbf{r}} = 0, \tag{2.13}$$

where we have used $\mathbf{r} \times \mathbf{r} = 0$. Now consider the derivative

$$\frac{d}{dt}(\mathbf{r} \times \dot{\mathbf{r}}) = \dot{\mathbf{r}} \times \dot{\mathbf{r}} + \mathbf{r} \times \ddot{\mathbf{r}} = \mathbf{r} \times \ddot{\mathbf{r}},$$

since $\dot{\mathbf{r}} \times \dot{\mathbf{r}} = 0$. Therefore, from equation (2.13) the derivative becomes

$$\frac{d}{dt}(\mathbf{r} \times \dot{\mathbf{r}}) = 0,$$

which we integrate to obtain

$$\mathbf{r} \times \dot{\mathbf{r}} = \mathbf{c} = constant. \tag{2.14}$$

Equation (2.14) is called the *area integral* and expresses the law of conservation of angular momentum for a unit mass. The vector $\mathbf{c}$ is called the *angular momentum vector*. Comparing $\mathbf{c}$ with the definition of an integral of the motion (§2.3),

$$\mathbf{c} = \mathbf{c}(\mathbf{r}(t), \dot{\mathbf{r}}(t)) = constant;$$

so we see that $\mathbf{c}$ indeed fulfills the definition of an integral.

An interesting fact about the geometry can be seen by taking the dot product of $\mathbf{c}$ with $\mathbf{r}$ and $\dot{\mathbf{r}}$:

$$\mathbf{r} \cdot \mathbf{c} = \mathbf{r} \cdot (\mathbf{r} \times \dot{\mathbf{r}}) = 0$$

and

$$\dot{\mathbf{r}} \cdot \mathbf{c} = \dot{\mathbf{r}} \cdot (\mathbf{r} \times \dot{\mathbf{r}}) = 0.$$

This result shows that $\mathbf{c}$ is normal to both $\mathbf{r}$ and $\dot{\mathbf{r}}$, and is therefore normal to the orbital plane of motion. We show this geometry in figure 2.3 in a cartesian coordinate system (x_1, x_2, x_3). Note that

$$\mathbf{r} \cdot \mathbf{c} = x_1 c_1 + x_2 c_2 + x_3 c_3 = 0$$

is the equation of a plane containing the origin (m_1 or m_2).

2.4.2 Energy Integral h

The derivation of the energy integral starts with taking the dot product of the velocity $\dot{\mathbf{r}}$ with equation (2.12):

$$\dot{\mathbf{r}} \cdot \left(\ddot{\mathbf{r}} + \frac{\mu}{r^3}\mathbf{r}\right) = 0. \tag{2.15}$$

Recall that we arrived at (2.14) by showing that (2.13) was a total time derivative. The same procedure will be used here. The first term is rather straightforward:

$$\dot{\mathbf{r}} \cdot \ddot{\mathbf{r}} = \frac{1}{2}\frac{d}{dt}(\dot{\mathbf{r}} \cdot \dot{\mathbf{r}}).$$

The second term is less obvious, but note that the total derivative

$$\frac{d}{dt}\left(\frac{-\mu}{\sqrt{\mathbf{r} \cdot \mathbf{r}}}\right) = \frac{\mu}{2\sqrt{(\mathbf{r} \cdot \mathbf{r})^3}}\frac{d}{dt}(\mathbf{r} \cdot \mathbf{r}) = \frac{\mu}{r^3}\dot{\mathbf{r}} \cdot \mathbf{r}.$$

Combine these two results by substituting them back into equation (2.15) to obtain

$$\frac{1}{2}\frac{d}{dt}(\dot{\mathbf{r}} \cdot \dot{\mathbf{r}}) + \frac{d}{dt}\left(-\frac{\mu}{\sqrt{\mathbf{r} \cdot \mathbf{r}}}\right) = 0.$$

The right side of this equation can be written

$$\frac{d}{dt}\left(\frac{1}{2}(\dot{\mathbf{r}} \cdot \dot{\mathbf{r}}) - \frac{\mu}{r}\right) = 0.$$

We immediately integrate this equation to obtain the energy integral

$$\frac{1}{2}(\dot{\mathbf{r}} \cdot \dot{\mathbf{r}}) - \frac{\mu}{r} = h = constant. \tag{2.16}$$

The constant h is called the energy because it is the sum of the kinetic and potential energies of the two-body system. Occasionally we will refer to h as the *two-body energy* or the *Keplerian energy*.

2.4.3 Laplacian Integral P

We take the cross product of the angular momentum equation (2.14) and equation (2.12) to obtain

$$\mathbf{c} \times \ddot{\mathbf{r}} + \frac{\mu}{r^3} (\mathbf{r} \times \dot{\mathbf{r}}) \times \mathbf{r} = 0.$$

Applying the elementary vector identity

$$(\mathbf{A} \times \mathbf{B}) \times \mathbf{C} = (\mathbf{A} \cdot \mathbf{C})\mathbf{B} - (\mathbf{B} \cdot \mathbf{C})\mathbf{A},$$

we obtain the result

$$\mathbf{c} \times \ddot{\mathbf{r}} + \frac{\mu}{r^3} [(\mathbf{r} \cdot \mathbf{r})\dot{\mathbf{r}} - (\dot{\mathbf{r}} \cdot \mathbf{r})\mathbf{r}] = 0. \tag{2.17}$$

Again we show that this equation can be written as a total derivative of another equation that we can integrate. The first term is simple, since $\mathbf{c}$ is constant:

$$\mathbf{c} \times \ddot{\mathbf{r}} = \frac{d}{dt}(\mathbf{c} \times \dot{\mathbf{r}}).$$

And again the second term is not as obvious; but note that

$$\mu \frac{d}{dt}\left(\frac{1}{r}\mathbf{r}\right) = \frac{\mu}{r^3}[(\mathbf{r} \cdot \mathbf{r})\dot{\mathbf{r}} - (\dot{\mathbf{r}} \cdot \mathbf{r})\mathbf{r}].$$

The right side of this equation is the same as the second term in (2.17). Substitute both these results into (2.17) to obtain

$$\frac{d}{dt}\left(\mathbf{c} \times \dot{\mathbf{r}} + \frac{\mu}{r}\mathbf{r}\right) = 0.$$

Integration results in

$$\mathbf{c} \times \dot{\mathbf{r}} + \frac{\mu}{r}\mathbf{r} = -\mathbf{P}. \tag{2.18}$$

This equation is called the *Laplacian integral*. $\mathbf{P}$ is a vector constant called the *Laplace vector* (also called the *eccentricity vector*). Note that we can use (2.14) to rewrite (2.18) as

$$(\mathbf{r} \times \dot{\mathbf{r}}) \times \dot{\mathbf{r}} + \frac{\mu}{r}\mathbf{r} = -\mathbf{P} \tag{2.19}$$

2.4.4 Relationship among the Integrals

The first relationship we show is that $\mathbf{c}$ is normal to $\mathbf{P}$. This can be seen by taking the dot product of (2.18) and $\mathbf{c}$:

$$\mathbf{c} \cdot \mathbf{c} \times \dot{\mathbf{r}} + \frac{\mu}{r}\mathbf{c} \cdot \mathbf{r} = -\mathbf{c} \cdot \mathbf{P}.$$

Since $\mathbf{c} \cdot \mathbf{r} = 0$ and $\mathbf{c} \cdot \mathbf{c} \times \dot{\mathbf{r}} = 0$ we get

$$\mathbf{c} \cdot \mathbf{P} = 0. \qquad (2.20)$$

Since $\mathbf{P}$ is normal to $\mathbf{c}$ and $\mathbf{c}$ in turn is normal to the orbital plane, then $\mathbf{P}$ must lie in the orbital plane (fig. 2.4). In this figure we show for the first time the angles Ω, i, and ω. In Appendix A we develop the relationship of $\hat{\mathbf{c}}$ and $\hat{\mathbf{P}}$ to these angles.

Another relationship among $\mathbf{c}$, h, and $\mathbf{P}$ can be found by taking the dot product of $\mathbf{P}$ with itself:

$$P^2 = (\mathbf{c} \times \dot{\mathbf{r}}) \cdot (\mathbf{c} \times \dot{\mathbf{r}}) + \frac{2\mu}{r}\mathbf{r} \cdot (\mathbf{c} \times \dot{\mathbf{r}}) + \frac{\mu^2}{r^2}(\mathbf{r} \cdot \mathbf{r}). \qquad (2.21)$$

Evaluate each term on the right side of (2.21). For the first term use the

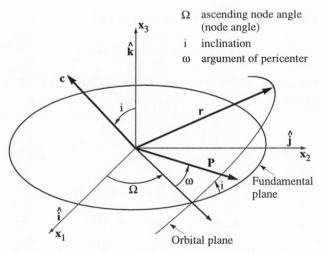

Figure 2.4   Geometry showing the Laplace vector ($\mathbf{P}$) and the angular momentum vector ($\mathbf{c}$).

vector identity

$$(\mathbf{A} \times \mathbf{B}) \cdot (\mathbf{C} \times \mathbf{D}) = (\mathbf{B} \cdot \mathbf{D})(\mathbf{C} \cdot \mathbf{A}) - (\mathbf{B} \cdot \mathbf{C})(\mathbf{D} \cdot \mathbf{A})$$

to arrive at

$$(\mathbf{c} \times \dot{\mathbf{r}}) \cdot (\mathbf{c} \times \dot{\mathbf{r}}) = (\dot{\mathbf{r}} \cdot \dot{\mathbf{r}})(\mathbf{c} \cdot \mathbf{c}) - (\mathbf{c} \cdot \dot{\mathbf{r}})(\mathbf{c} \cdot \dot{\mathbf{r}}).$$

Note that $\mathbf{c} \cdot \dot{\mathbf{r}} = 0$ and $\mathbf{c} \cdot \mathbf{c} = c^2$, so the first term becomes

$$(\mathbf{c} \times \dot{\mathbf{r}}) \cdot (\mathbf{c} \times \dot{\mathbf{r}}) = c^2 \dot{\mathbf{r}} \cdot \dot{\mathbf{r}}.$$

For the second term we use another vector identity,

$$\mathbf{A} \cdot \mathbf{B} \times \mathbf{C} = \mathbf{A} \times \mathbf{B} \cdot \mathbf{C},$$

to arrive at

$$\frac{2\mu}{r} \mathbf{r} \cdot (\mathbf{c} \times \dot{\mathbf{r}}) = -\frac{2\mu}{r} \mathbf{r} \times \dot{\mathbf{r}} \cdot \mathbf{c}$$

$$= -\frac{2\mu}{r} c^2.$$

The third term is

$$\frac{\mu^2}{r^2} (\mathbf{r} \cdot \mathbf{r}) = \mu^2.$$

Put these results back into equation (2.21) and we get

$$P^2 = c^2 \left(\dot{\mathbf{r}} \cdot \dot{\mathbf{r}} - \frac{2\mu}{r} \right) + \mu^2.$$

Note from equation (2.16) that the parenthetical expression in the above equation is just twice the energy. We therefore have

$$P^2 = 2hc^2 + \mu^2. \tag{2.22}$$

Rearrange (2.22) to put the equation in the form

$$\frac{c^2}{\mu} = -\frac{\mu}{2h} \left(1 - \frac{P^2}{\mu^2} \right). \tag{2.23}$$

Equations (2.20) and (2.23) show that the integrals of the motion given by equations (2.14), (2.16), and (2.19) are not independent of each other.

2.4.5 Summary of Integrals Derived

The equations (2.14), (2.16), and (2.18), which are

$$\mathbf{r} \times \dot{\mathbf{r}} = \mathbf{c}$$

$$\frac{1}{2}\dot{\mathbf{r}} \cdot \dot{\mathbf{r}} - \frac{\mu}{r} = h$$

$$\mathbf{c} \times \dot{\mathbf{r}} + \frac{\mu}{r}\mathbf{r} = -\mathbf{P},$$

provide a total of seven constants of the motion (Pollard [58]). However, the scalar relations we just derived (eqs. 2.20 and 2.23),

$$\mathbf{c} \cdot \mathbf{P} = 0$$

$$\frac{c^2}{\mu} = -\frac{\mu}{2h}\left(1 - \frac{P^2}{\mu^2}\right),$$

reduce the number of independent integrals found so far to five. Since we require six constants for a complete solution to the two-body problem, we still have one to find. We defer the solution for the final integral until after the following application.

2.4.6 Application: The Abort Problem

This solution (Bond [11]) was used to size rocket engines that were later installed on the Apollo spacecraft. This application demonstrates that from the knowledge of only two integrals (the energy and magnitude of the angular

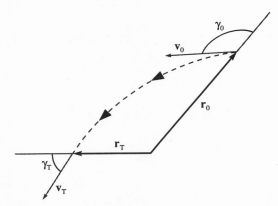

Figure 2.5 Abort problem.

momentum), the practical solution of an important spacecraft design problem can be found. The problem is as follows (fig. 2.5):

Given   The initial position magnitude (r_0), terminal position and velocity magnitudes (r_T, v_T), and flight path angle (γ_T).
Find   The initial velocity magnitude (v_0) and flight path angle (γ_0).
Procedure

▷ Compute v_0 from energy integral (eq. 2.16).

$$h = \frac{v_0^2}{2} - \frac{\mu}{r_0} = \frac{v_T^2}{2} - \frac{\mu}{r_T} = constant,$$

where $v_0^2 = \dot{\mathbf{r}}_0 \cdot \dot{\mathbf{r}}_0$ and $v_T^2 = \dot{\mathbf{r}}_T \cdot \dot{\mathbf{r}}_T$. The initial velocity is the only unknown in this equation, so

$$v_0 = \sqrt{2\left(\frac{v_T^2}{2} - \frac{\mu}{r_T} + \frac{\mu}{r_0}\right)}.$$

▷ Compute γ_0 from area integral. From (2.14),

$$c = \|\mathbf{r}_0 \times \mathbf{v}\| = \|\mathbf{r}_T \times \mathbf{v}_T\| = constant,$$

therefore,

$$c = r_0 v_0 \sin(\gamma_0) = r_T v_T \sin(\gamma_T) = constant,$$

where γ_0 and γ_T are the flight path angles measured from the local verticals. Having already found v_0 from step 1, we now solve for γ_0 to obtain

$$\gamma_0 = \sin^{-1}\left[\left(\frac{r_T v_T}{r_0 v_0} \sin(\gamma_T)\right)\right].$$

Note that there are two solutions for γ_0 (fig. 2.6), one in the first quadrant $(\gamma_0^{(1)})$ and one in the second quadrant $(\gamma_0^{(2)})$.

End.

2.4.7 Kepler's Equation

We now derive the sixth integral of the motion for the two-body problem. We start by taking the dot product of the angular momentum vector (2.14) with itself:

$$\mathbf{c} \cdot \mathbf{c} = c^2 = (\mathbf{r} \times \dot{\mathbf{r}}) \cdot (\mathbf{r} \times \dot{\mathbf{r}}),$$

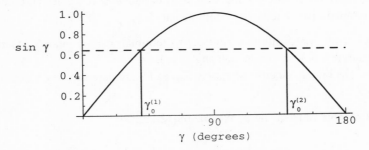

Figure 2.6 Two solutions for the flight path angle.

and using a vector identity already given above,

$$c^2 = (\dot{\mathbf{r}} \cdot \dot{\mathbf{r}})(\mathbf{r} \cdot \mathbf{r}) - (\dot{\mathbf{r}} \cdot \mathbf{r})(\mathbf{r} \cdot \dot{\mathbf{r}}). \tag{2.24}$$

Solve the energy integral (2.16) for $(\dot{\mathbf{r}} \cdot \dot{\mathbf{r}})$:

$$\dot{\mathbf{r}} \cdot \dot{\mathbf{r}} = 2h + \frac{2\mu}{r}.$$

Use the following elementary derivation to solve for $(\mathbf{r} \cdot \dot{\mathbf{r}})$:

$$\frac{d}{dt}(r^2) = \frac{d}{dt}(\mathbf{r} \cdot \mathbf{r}). \tag{2.25}$$

$$2r\dot{r} = (\mathbf{r} \cdot \dot{\mathbf{r}}) + (\dot{\mathbf{r}} \cdot \mathbf{r}) = 2(\mathbf{r} \cdot \dot{\mathbf{r}}). \tag{2.26}$$

$$r\dot{r} = \mathbf{r} \cdot \dot{\mathbf{r}}. \tag{2.27}$$

Substituting all this into (2.24) results in

$$c^2 = r^2 \left(2h + \frac{2\mu}{r} \right) - r^2 \dot{r}^2. \tag{2.28}$$

We will reduce equation (2.25) to the form

$$dt = f(r)dr,$$

which we will integrate to get an expression for time and a constant of integration.

Solving (2.25) for $r\dot{r}$ we obtain

$$r\dot{r} = \pm\sqrt{-2h} \sqrt{-\left(r^2 + \frac{\mu r}{h} \right) + \frac{c^2}{2h}}. \tag{2.29}$$

At this point we will assume the energy (h) is negative.[1] We consider the case for $h > 0$ in Appendix B.

Continuing with (2.26), we next complete the square inside the braces:

$$r\dot{r} = \pm\sqrt{-2h}\sqrt{-\left(r^2 + \frac{\mu r}{h} + \left(\frac{\mu}{2h}\right)^2\right) + \frac{c^2}{2h} + \left(\frac{\mu}{2h}\right)^2}$$

$$= \pm\sqrt{-2h}\sqrt{-\left(r + \frac{\mu}{2h}\right)^2 + \frac{c^2}{2h} + \left(\frac{\mu}{2h}\right)^2}. \tag{2.30}$$

Now multiply equation (2.23) by the factor $\mu/2h$ to simplify the right side of equation (2.27),

$$\frac{c^2}{2h} = -\left(\frac{\mu}{2h}\right)^2\left(1 - \frac{P^2}{\mu^2}\right)$$

or

$$\frac{c^2}{2h} + \left(\frac{\mu}{2h}\right)^2 = \left(\frac{P}{2h}\right)^2.$$

Therefore, equation (2.27) becomes

$$r\frac{dr}{dt} = \pm\sqrt{-2h}\sqrt{\left(\frac{P}{2h}\right)^2 - \left(r + \frac{\mu}{2h}\right)^2},$$

which we now write as

$$\pm\sqrt{-2h}\,dt = \frac{r\,dr}{\sqrt{\left(\frac{P}{2h}\right)^2 - \left(r + \frac{\mu}{2h}\right)^2}}, \tag{2.31}$$

which is the desired form.

The denominator of equation (2.28) is rather unwieldy but we can reduce the right side to recognizable integrals by the substitution,

$$z = r + \frac{\mu}{2h}, \qquad dz = dr, \tag{2.32}$$

to obtain

$$\pm\sqrt{-2h}\,dt = \frac{z - (\mu/2h)}{\sqrt{\left(\frac{P}{2h}\right)^2 - z^2}}\,dz.$$

[1] The sign of the energy (negative, zero or positive) will be used to identify the type of orbit, which can be either elliptical $(h < 0)$, hyperbolic $(h > 0)$, or parabolic $(h = 0)$. We are for now assuming that the orbit is elliptical.

From the integral tables,

$$\pm\sqrt{-2h}\ (t + \text{constant}) = -\sqrt{\left(\frac{P}{2h}\right)^2 - z^2}$$

$$-\frac{\mu}{2h}\left[-\cos^{-1}\left(\frac{z}{P/2h}\right)\right]. \tag{2.33}$$

By defining the angle E as

$$E = \cos^{-1}\left(\frac{z}{P/2h}\right) \tag{2.34}$$

and since

$$\sin^2 E = 1 - \cos^2 E,$$

we obtain

$$\sin E = \frac{2h}{P}\sqrt{\left(\frac{P}{2h}\right)^2 - z^2}. \tag{2.35}$$

Now substitute equations (2.31) and (2.32) into equation (2.30) to obtain

$$\pm\sqrt{-2h}\ (t + \text{constant}) = -\frac{P}{2h}\sin E + \frac{\mu}{2h}E.$$

Multiply through by $2h/\mu$ and reorganize to arrive at Kepler's equation,

$$\pm\frac{2h}{\mu}\sqrt{-2h}\ (t + K) = E - \frac{P}{\mu}\sin E, \tag{2.36}$$

where K is the integration constant.

Since equation (2.33) can be rearranged to have only K on the left side and all the other terms on the right side, K then is an integral of the motion in the sense of §2.2 and is the sixth integral required for the complete solution of the two-body problem. It must be stressed that equation (2.33) applies only to orbits for which $h < 0$.

Finally, observe that an expression for $r = r(E)$ is straightforward from equations (2.32) and (2.34):

$$r = z - \frac{\mu}{2h}$$

$$= \frac{P}{2h}\cos E - \frac{\mu}{2h}$$

$$r = -\frac{\mu}{2h}\left[1 - \frac{P}{\mu}\cos E\right]. \tag{2.37}$$

3

KEPLER'S LAWS

The story of Johannes Kepler (1571–1630) and the discovery of the three empirical laws that bear his name is well known. The story has been blended into a sort of scientific legend. Tales of his interaction with Tycho Brahe, for instance, have kept students amused for countless number of class hours down through the years. The legends vary between entertaining yarns and outright sensationalism and unfortunately tend to obscure the real man and his work. We recommend the paper by Otto Volk [76] as a starting point for anyone interested in seriously delving into Kepler's life. A list of the main references in Volk's paper was compiled and published "on the Occasion of Kepler's 400th Birthday (December 27, 1571, Julian Calendar)."

In addition to the discovery of the three laws, Volk cites other important scientific and mathematical contributions by Kepler:

1. A theory of optics, including an invention of a telescope.
2. Work in infinitesimal calculus.
3. Work in the use of logarithms in astronomical calculations.
4. Contributions to the theory of conic sections.
5. Work in polygons and polyhedra.

The three laws of Kepler are given by (Szebehely [70]):

1. The orbits of the planets are ellipses with the Sun at the focus.
2. The vector connecting the Sun and a planet sweeps out equal areas in equal time.
3. The square of the periods of the planets are proportional to the cubes of their semi-major axes.

The first two of these laws were published in 1609 and the third law was published in 1618. Not all authors agree on the order of Kepler's first and second laws. However, the order in which they are given above seems to be preferred by most authors; see Bate *et al.* [3], Battin [4], Danby [28], Green

[38], Pollard [58], Szebehely [70], and Taff [72]. They are given in opposite order by Brouwer and Clemence [24] and Moulton [54].

In this chapter we will verify the laws of Kepler by manipulation of the integrals derived in chapter 2.

3.1 Kepler's First Law

We begin this derivation by taking the dot product of the position $\mathbf{r}$ and the Laplacian integral (eq. 2.18),

$$\mathbf{r} \cdot \mathbf{c} \times \dot{\mathbf{r}} + \frac{\mu}{r}\mathbf{r} \cdot \mathbf{r} = -\mathbf{r} \cdot \mathbf{P}. \tag{3.1}$$

The first term on the left side reduces to

$$\mathbf{r} \cdot \mathbf{c} \times \dot{\mathbf{r}} = -\mathbf{c} \cdot \mathbf{r} \times \dot{\mathbf{r}} = -c^2.$$

The second term on the left side reduces at once to μr. The right side of equation (3.1) becomes

$$\mathbf{r} \cdot \mathbf{P} = rP \cos\phi,$$

where we have defined the angle ϕ to be the angle between the position vector $\mathbf{r}$ and the Laplace vector $\mathbf{P}$. Substitute these back into (3.1) to get

$$-c^2 + \mu r = -rP \cos\phi.$$

Now solve for the distance r (which is the magnitude of the position vector $\mathbf{r}$), which gives

$$r = \frac{(c^2/\mu)}{[1 + (P/\mu)\cos\phi]}. \tag{3.2}$$

Comparing this equation to the standard equation of a conic section, (Appendix C), which is

$$r = \frac{p}{[1 + e\cos\phi]}, \tag{3.3}$$

where

$$p \equiv a(1 - e^2),$$

we obtain

$$p = \frac{c^2}{\mu}, \tag{3.4}$$

which is known as the *semi-latus rectum*, and

$$e = \frac{P}{\mu},$$

which is the *eccentricity*, and a is the *semi-major axis* of the conic.

Using equation (2.23) and the above definitions, we can relate the energy constant with one of the conic parameters,

$$\frac{c^2}{\mu} = -\frac{\mu}{2h}(1 - (P/\mu)^2) = p \equiv a(1 - e^2).$$

Therefore,

$$a = -\frac{\mu}{2h}, \tag{3.5}$$

which states the important result that the semi-major axis of the conic depends *only* on the energy.

Note that when $\phi = 0$, $\cos\phi = 1$ and the distance r (from eq. 3.3) assumes its minimum value, which is defined as

$$r_\pi = \frac{p}{(1 + e)} = \frac{a(1 - e^2)}{(1 + e)} = a(1 - e).$$

r_π is called the *pericenter distance*: "peri" means near, so r_π is the nearest distance from the orbit to the center of attraction, which of course is either mass of the two-body system. The subscript π means that any quantity to which it is attached is evaluated at $\phi = 0$.

We see that by defining the angle ϕ to be between the Laplace vector **P** and the position vector **r**, we orient **P** to lie along the line between the center of attraction and the pericenter point (fig. 3.1). The angle ϕ is called the *true anomaly*.

Equation (3.2), which we recall was derived only from Newton's laws, is a verification of Kepler's first law. A generalization of this law is that the orbit of m_2 (or m_1) is a conic section with m_1 (or m_2) at one of the foci (called the *prime focus*). Note that this verification was done without any assumption about the energy. This result is valid for any conic section—ellipse ($h < 0$), hyperbola ($h > 0$), or parabola ($h = 0$).

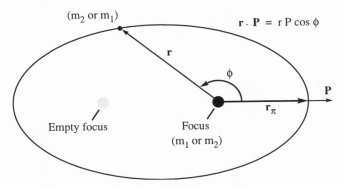

Figure 3.1   The elliptical conic section.

3.2 Kepler's Second Law

This verification begins with the area integral, equation (2.14). For convenience we express the area integral in a coordinate system defined by the local vertical ($\hat{\mathbf{r}}$) and local horizontal ($\hat{\boldsymbol{\phi}}$) unit vectors.

In polar coordinates the position and velocity vectors (fig. 3.2) are

$$\mathbf{r} = r\,\hat{\mathbf{r}} \qquad\qquad (3.6)$$

$$\dot{\mathbf{r}} = \dot{r}\,\hat{\mathbf{r}} + r\dot{\phi}\,\hat{\boldsymbol{\phi}},$$

where the unit vector $\hat{\boldsymbol{\phi}}$, which is in the direction of the local horizontal, is

$$\hat{\boldsymbol{\phi}} = \hat{\mathbf{c}} \times \hat{\mathbf{r}},$$

and the unit vector perpendicular to the orbit plane, $\hat{\mathbf{c}}$, is defined by

$$\hat{\mathbf{c}} = \frac{1}{c}\,\mathbf{c}.$$

The unit vectors $\hat{\mathbf{r}}$, $\hat{\boldsymbol{\phi}}$, $\hat{\mathbf{c}}$ form an orthogonal system. The angular momentum $\hat{\mathbf{c}}$ in this system is

$$\mathbf{c} = \mathbf{r} \times \dot{\mathbf{r}} = r\,\hat{\mathbf{r}} \times (\dot{r}\,\hat{\mathbf{r}} + r\dot{\phi}\hat{\boldsymbol{\phi}}).$$

Since $\mathbf{r} \times \mathbf{r} = 0$ and $\hat{\mathbf{r}} \times \hat{\boldsymbol{\phi}} = \hat{\mathbf{c}}$, we get

$$\mathbf{c} = r^2\dot{\phi}\,\hat{\mathbf{c}},$$

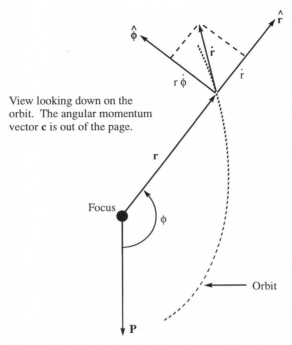

View looking down on the
orbit. The angular momentum
vector **c** is out of the page.

Figure 3.2   The local vertical, local horizontal system.

so that the magnitude of the angular momentum is

$$c = r^2\dot{\phi} = \text{constant.} \tag{3.7}$$

The rate at which area is swept out by the distance in a conic section is

$$\frac{dA}{dt} = \frac{1}{2}r^2\frac{d\phi}{dt},$$

which comes from taking the limit of the incremental area

$$\Delta A = \frac{1}{2}\ base\ \times\ height$$

$$= \frac{1}{2}(r + \Delta r)(r\,\Delta\phi)$$

or, more directly,

$$dA = \lim_{\Delta r \to 0}\Delta A = \frac{1}{2}r^2 d\phi.$$

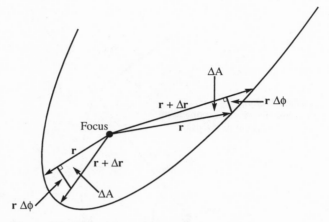

Figure 3.3   Sectorial area of a conic.

Comparing this equation with equation (3.7), we see that

$$\frac{dA}{dt} = \frac{1}{2}c = constant.$$

This equation is the mathematical statement of Kepler's second law: The line joining the planet to the Sun sweeps out equal areas in equal time.

3.3 Kepler's Third Law

This law can be verified from equation (2.28), which we introduced as an intermediate step in the development of Kepler's equation.

Now recall that we are considering the case for which $h < 0$. From the discussion of Kepler's first law, we introduced the semi-major axis and eccentricity

$$a = -\frac{\mu}{2h}$$

$$e = \frac{P}{\mu},$$

so equation (2.37), which is

$$r = -\frac{\mu}{2h}\left[1 - \frac{P}{\mu}\cos E\right],$$

becomes

$$r = a(1 - e \cos E).\tag{3.8}$$

Using this result, equation (2.31) becomes

$$\sqrt{\frac{\mu}{a}}\, dt = \frac{a(1 - e \cos E)(ae \sin E)dE}{\sqrt{a^2 e^2 - a^2 e^2 \cos^2 E}},$$

which after the cancelations becomes

$$\sqrt{\frac{\mu}{a}}\, dt = a(1 - e \cos E)dE.\tag{3.9}$$

Now rearrange equation (3.9) slightly and integrate

$$\int_0^T dt = \sqrt{\frac{a^3}{\mu}}\int_0^{2\pi} (1 - e \cos E)dE,\tag{3.10}$$

where T is one orbital period, or the time for E to increase from 0 to 2π. Performing the integration we obtain

$$T = 2\pi\sqrt{\frac{a^3}{\mu}}.\tag{3.11}$$

Now square both sides:

$$T^2 = \left(\frac{4\pi^2}{\mu}\right)a^3,\tag{3.12}$$

which is a verification of Kepler's third law: The square of the period of an orbit is proportional to the cube of the orbit's semi-major axis. Note that this law applies only to the ellipse.

3.3.1 Planetary Mass Determination

Kepler's third law can be used directly in the determination of the ratio of the mass of a planet to the mass of the Sun. Below we will discuss two approaches to this application. In the first approach we will use the (observed) period and the semi-major axis of both a planet and an asteroid of negligible mass revolving about the Sun. The observed quantities are as follows:

T = period of the planet
a = semi-major axis of the planet's orbit

T_1 = period of the asteroid

a_1 = semi-major axis of the asteroid's orbit.

Using Kepler's third law (eq. 3.12) for the two-body motion of the Sun and planet, we have

$$T^2 = \frac{4\pi^2}{G(S+P)} a^3, \tag{3.13}$$

where

S = mass of the Sun

P = mass of the planet

G = Universal Gravitational Constant.

Note that for this case the factor μ as defined in chapter 2 is $G(S+P)$. Similarly, we can write Kepler's third law for the two-body motion of the Sun and the asteroid,

$$T_1^2 = \frac{4\pi^2}{G(S+m_1)} a_1^3, \tag{3.14}$$

where

$$m_1 = \text{mass of the asteroid} \ll S.$$

The assumption $m_1 \ll S$ permits us to ignore the negligible mass of the asteroid. Now take the ratio of equation (3.13) to (3.14),

$$\left(\frac{T}{T_1}\right)^2 = \frac{a^3}{G(S+P)} \frac{G(S+m_1)}{a_1^3}$$

$$= \frac{GS(1+m_1/S)}{GS(1+P/S)} \left(\frac{a}{a_1}\right)^3.$$

With the assumption $m_1 \ll S$ we can solve directly for the ratio

$$\frac{P}{S} = \left(\frac{a}{a_1}\right)^3 \left(\frac{T_1}{T}\right)^2 - 1. \tag{3.15}$$

Since T, a, T_1, and a_1 are all known by observation and the mass of the asteroid as negligible, we can solve for the ratio of the planet's mass to the Sun's mass.

In the second approach we will again use the observer values of the period and semi-major axis of the planetary orbit about the Sun; but in this approach we use observed values of a satellite about the planet. Let

T = period of the planet

a = semi-major axis of the planet's orbit

M = mass of the planet

T_s = period of the satellite

a_s = semi-major axis of the satellite's orbit

m_s = mass of the satellite.

Now we write Kepler's third law for the motion of the satellite about the planet,

$$T_s^2 = \frac{4\pi^2}{G(M + m_s)} a_s^3. \tag{3.16}$$

In equation (3.13) we set

$$P = M + m_s.$$

This way we consider the planet mass and the mass of its satellite to be one mass orbiting the Sun. Equation (3.13) becomes

$$T^2 = \frac{4\pi^2}{G(S + M + m_s)} a^3. \tag{3.17}$$

Now take the ratio of equation (3.16) to equation (3.17), obtaining

$$\left(\frac{T_s}{T}\right)^2 = \frac{a_s^3}{G(M + m_s)} \frac{G(S + M + m_s)}{a^3}$$

$$= \frac{G}{G} \left(\frac{S}{M + m_s} + 1\right) \left(\frac{a_s}{a}\right)^3.$$

Now solve for the ratio of the mass of the Sun to the combined mass of the planet and satellite to obtain

$$\frac{S}{M + m_s} = \left(\frac{a}{a_s}\right)^3 \left(\frac{T_s}{T}\right)^2 - 1. \tag{3.18}$$

We take this opportunity to point out that in celestial and orbital mechanics we use mass ratios in our calculations. The reason is that mass ratios such as in equations (3.15) and (3.18) can be very accurately determined from observations, but G, the Universal Gravitational Constant, is poorly known (to about five digits). Refer back to §1.2, where we gave an example calculating the gravitational constant for Jupiter (GJ) from the Sun/Jupiter mass ratio and the heliocentric gravitational constant (GS).

4

METHODS OF COMPUTATION

4.1 Position/Velocity in Integrals

In the practical application of orbital mechanics to the analysis of the orbits of satellites and spacecraft, we usually want to know the solution of the two-body problem in a system of coordinates rather than the integrals of motion. Most often the solution is required to be in the form of the position and velocity in the cartesian coordinates of an inertial reference system.

In this chapter we develop a solution for position and velocity ($\mathbf{r}$ and $\dot{\mathbf{r}}$) as a function of the time for the elliptical ($h < 0$) case. We begin with the integrals of the motion derived in chapter 2 and relate them to the geometry discussed in chapter 3. The six integrals of the motion of the two-body problem were shown in chapter 2 to be (eqs. 2.14, 2.16, 2.19, 2.36, and 2.37)

$$\mathbf{r} \times \dot{\mathbf{r}} = \mathbf{c}$$

$$\frac{1}{2}\dot{\mathbf{r}} \cdot \dot{\mathbf{r}} - \frac{\mu}{r} = h$$

$$(\mathbf{r} \times \dot{\mathbf{r}}) \times \dot{\mathbf{r}} + \frac{\mu}{r}\mathbf{r} = -\mathbf{P}$$

$$\pm \frac{2h}{\mu}\sqrt{-2h}\,(t + K) = E - \frac{P}{\mu}\sin E,$$

where

$$r = -\frac{\mu}{2h}\left(1 - \frac{P}{\mu}\cos E\right).$$

Equations (2.36) for the time and (2.37) for the distance are valid only for elliptic motion $(h < 0)$.

To reduce the eight constants ($\mathbf{c}$, h, $\mathbf{P}$, and K) to six, we employed the two scalar relations (eqs. 2.20 and 2.23),

$$\mathbf{c} \cdot \mathbf{P} = 0$$

$$\frac{c^2}{\mu} = -\frac{\mu}{2h}(1 - (P/\mu)^2).$$

In chapter 3 we also identified certain conic parameters with some of the integrals (eqs. 3.4 and 3.5):

$$p \equiv a(1 - e^2) = \frac{c^2}{\mu}$$

$$e = \frac{P}{\mu}$$

$$a = -\frac{\mu}{2h},$$

as well as the conic relation for the distance (eq. 3.3),

$$r = \frac{p}{[1 + e \cos \phi]}.$$

Note that equation (2.36) (Kepler's equation) can now be written

$$n(t - t_\pi) = M = E - e \sin E, \tag{4.1}$$

where we have used the boundary condition

$$K = -t_\pi \tag{4.2}$$

when $E = 0$. We have also introduced

$$n = -\frac{2h}{\mu}\sqrt{-2h} = \sqrt{\frac{\mu}{a^3}} \tag{4.3}$$

and

$$M = n(t - t_\pi), \tag{4.4}$$

where t_π is called the *time of pericenter passage*, n is called the *mean motion*, and M is called the *mean anomaly*. Equation (2.37) for the distance can be written as

$$r = a(1 - e \cos E). \tag{4.5}$$

The angle E is called the *eccentric anomaly*.

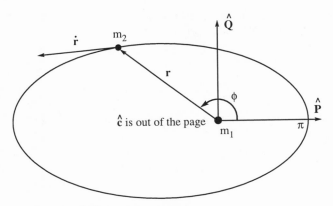

Figure 4.1 $\hat{\mathbf{P}}, \hat{\mathbf{Q}}, \hat{\mathbf{c}}$ coordinate system.

4.2 Position/Velocity—True Anomaly

Consider the two-body motion of the mass m_2 with respect to the mass m_1 in an inertial cartesian coordinate system defined by a fixed origin at the mass m_1. The fundamental plane of the system is the plane normal to the angular momentum vector $\mathbf{c}$, and the principal direction is along the Laplace vector $\mathbf{P}$ (fig. 4.1). Let

$$\hat{\mathbf{P}} = \frac{1}{P}\mathbf{P}$$

$$\hat{\mathbf{c}} = \frac{1}{c}\mathbf{c}$$

$$\hat{\mathbf{Q}} = \hat{\mathbf{c}} \times \hat{\mathbf{P}}.$$

The position vector in terms of ϕ is

$$\mathbf{r} = r\cos\phi\,\hat{\mathbf{P}} + r\sin\phi\,\hat{\mathbf{Q}}. \tag{4.6}$$

To obtain the velocity $\dot{\mathbf{r}}$ we differentiate equation (4.6):

$$\dot{\mathbf{r}} = (\dot{r}\cos\phi - r\sin\phi\dot{\phi})\hat{\mathbf{P}} + (\dot{r}\sin\phi + r\cos\phi\dot{\phi})\hat{\mathbf{Q}}.$$

We get $\dot{\phi}$ from our discussion of Kepler's second law (eq. 3.7):

$$r^2\dot{\phi} = c = \sqrt{\mu p}. \tag{4.7}$$

We get $\dot{r}$ by differentiating the conic equation (3.3):

$$r = \frac{p}{[1 + e\cos\phi]}$$

$$\dot{r} = \left[\frac{-p}{(1 + e\cos\phi)^2}\right](-e\sin\phi\dot{\phi})$$

$$= \left(\frac{e\sin\phi}{p}\right)r^2\dot{\phi}.$$

Using (4.7),

$$\dot{r} = e\sqrt{\frac{\mu}{p}}\ \sin\phi. \tag{4.8}$$

Now substitute equations (4.7) and (4.8) for $\dot{\phi}$ and $\dot{r}$ into the expression for $\dot{\mathbf{r}}$,

$$\dot{\mathbf{r}} = \left[\left(e\sqrt{\frac{\mu}{p}}\sin\phi\right)\cos\phi - \frac{r\sin\phi\sqrt{\mu p}}{r^2}\right]\hat{\mathbf{P}}$$

$$+ \left[\left(e\sqrt{\frac{\mu}{p}}\sin\phi\right)\sin\phi + \frac{r\cos\phi\sqrt{\mu p}}{r^2}\right]\hat{\mathbf{Q}}$$

$$= \sqrt{\frac{\mu}{p}}\left[e\sin\phi\cos\phi - \sin\phi\frac{p}{r}\right]\hat{\mathbf{P}} + \sqrt{\frac{\mu}{p}}\left[e\sin^2\phi + \cos\phi\frac{p}{r}\right]\hat{\mathbf{Q}}.$$

Now, from equation (3.3),

$$\frac{p}{r} = 1 + e\cos\phi,$$

which we substitute into the equation for $\dot{\mathbf{r}}$ to obtain

$$\dot{\mathbf{r}} = \sqrt{\frac{\mu}{p}}\ [e\sin\phi\cos\phi - \sin\phi(1 + e\cos\phi)]\hat{\mathbf{P}}$$

$$+ \sqrt{\frac{\mu}{p}}\ [e\sin^2\phi + \cos\phi(1 + e\cos\phi)]\hat{\mathbf{Q}}.$$

Making the cancelations we finally get

$$\dot{\mathbf{r}} = \sqrt{\frac{\mu}{p}}\ \left[-\sin\phi\ \hat{\mathbf{P}} + (e + \cos\phi)\ \hat{\mathbf{Q}}\right]. \tag{4.9}$$

Equations (4.6) for $\mathbf{r}$ and (4.9) for $\dot{\mathbf{r}}$ give expressions for position and velocity in terms of true anomaly. This is *not* a solution because it does not give the solution as function of time. The next step in obtaining a solution is to eliminate the true anomaly ϕ in favor of eccentric anomaly E which is related to time through Kepler's equation, equation (4.1).

4.3 Position/Velocity—Eccentric Anomaly

Equations (4.6) and (4.9) have terms involving both $\cos\phi$ and $\sin\phi$. We now solve for these functions in terms of $\cos E$ and $\sin E$.

We have two equations for the distance:

$$r = \frac{a(1 - e^2)}{1 + e\cos\phi} = a(1 - e\cos E).$$

Solving for $\cos\phi$,

$$\cos\phi = \frac{\cos E - e}{1 - e\cos E}. \tag{4.10}$$

We also have equation (4.8) for $\dot{r}$ in terms of $\sin\phi$:

$$\dot{r} = e\sqrt{\frac{\mu}{p}}\,\sin\phi.$$

We can also get an expression for $\dot{r}$ in terms of $\sin E$. From equation (4.5):

$$r = a(1 - e\cos E).$$

Differentiate to get $\dot{r}$:

$$\dot{r} = ae\sin E\,\dot{E}. \tag{4.11}$$

From equation (3.9), which we used to verify Kepler's third law,

$$\frac{dE}{dt} = \frac{\sqrt{\mu/a}}{a(1 - e\cos E)} = \frac{\sqrt{\mu/a}}{r}. \tag{4.12}$$

Substitute (4.12) into (4.11) to obtain

$$\dot{r} = \frac{e\sqrt{\mu/a}\,\sin E}{1 - e\cos E}. \tag{4.13}$$

Note that equation (4.13) can also be written

$$r\dot{r} = e\sqrt{\mu a}\sin E$$

by the use of equation (4.5).

Now equate equations (4.8) and (4.13) for $\dot{r}$ and solve for $\sin \phi$:

$$\sin \phi = \frac{\sqrt{p/a}\ \sin E}{1 - e \cos E}.$$

Since $p \equiv a(1 - e^2)$,

$$\sin \phi = \frac{\sqrt{1 - e^2}\ \sin E}{1 - e \cos E}. \tag{4.14}$$

Now substitute equations (4.10) and (4.14) into equation (4.6):

$$\mathbf{r} = r \cos \phi\, \hat{\mathbf{P}} + r \sin \phi\, \hat{\mathbf{Q}}$$

$$= r \left[\frac{\cos E - e}{1 - e \cos E} \right] \hat{\mathbf{P}} + r \left[\frac{\sqrt{1 - e^2}\, \sin E}{1 - e \cos E} \right] \hat{\mathbf{Q}}.$$

Recalling that $r = a(1 - e \cos E)$ and $p \equiv a(1 - e^2)$, we have finally:

$$\mathbf{r} = a(\cos E - e)\, \hat{\mathbf{P}} + \sqrt{ap}\, \sin E\, \hat{\mathbf{Q}}. \tag{4.15}$$

To obtain $\dot{\mathbf{r}}$ we differentiate (4.15) to obtain

$$\dot{\mathbf{r}} = -a \sin E\, \dot{E}\, \hat{\mathbf{P}} + \sqrt{ap}\, \cos E\, \dot{E}\, \hat{\mathbf{Q}}.$$

Using the equation for $\dot{E}$ (4.12), we arrive at

$$\dot{\mathbf{r}} = -\frac{\sqrt{\mu a}}{r} \sin E\, \hat{\mathbf{P}} + \frac{\sqrt{\mu p}}{r} \cos E\, \hat{\mathbf{Q}}. \tag{4.16}$$

Equations (4.15) and (4.16) give expressions for $\mathbf{r}$ and $\dot{\mathbf{r}}$ in terms of E, the Eccentric Anomaly. Since E is related to t through Kepler's equation (eq. 4.1), we have a solution, $\mathbf{r}(t)$ and $\dot{\mathbf{r}}(t)$, for the elliptical case.

4.4 Solution of Kepler's Equation

Kepler's equation can be solved by a Newton-Raphson method, which is a numerical technique for finding roots of an equation. Using equation (4.1), define the function $F(E)$ to be

$$F(E) = E - e \sin E - n(t - t_\pi). \tag{4.17}$$

All quantities in this equation are known *except* E. We want to find the value of E for which $F(E) = 0$, thus satisfying Kepler's equation. A convenient

Table 4.1   Newton-Raphson convergence for calculating the eccentric anomaly.

Iteration Number, k	$|E - E_k|$	E (degrees)
1	0.9341957	87.69266
2	0.2415711	73.85166
3	0.0251371	72.41141
4	0.0002688	72.39601
5	10^{-7}	72.39601

iterative solution can be found by expanding $F(E)$ in a Taylor series about some approximate value E_k:

$$F(E) = F(E_k) + \left(\frac{dF}{dE}\right)_k (E - E_k) + \dots$$

Solve for E recalling the condition that $F(E) = 0$ to obtain

$$E = E_k - \frac{F(E_k)}{\left(\frac{dF}{dE}\right)_k}. \tag{4.18}$$

Now solve for $(\frac{dF}{dE})_k$ by using equation (4.17):

$$\left(\frac{dF}{dE}\right)_k = 1 - e\cos E_k. \tag{4.19}$$

Equation (4.18) only provides an approximation for E. We then successively calculate more accurate values for E by using (4.18) with the current value of E_k set to the value for E found from the previous iteration. We end the iterative process when the values for E and E_k agree to some desired accuracy. A numerical example of this procedure follows. Consider the geocentric orbit defined by

$a = 24{,}400$ km
$e = 0.7$
$t = 3600$ seconds
$t_\pi = 0$
tolerance $= 10^{-6}$
$\mu = 398601$ km$^3 \cdot$ sec^{-2}
$E_1 = \sqrt{\frac{\mu}{a^3}} (t - t_\pi)(=$ first guess$)$.

The iteration procedure based on equation (4.18) converges in five iterations, as shown by table 4.1.

4.5 Computation of Position/Velocity

In this section we present algorithms for computing the position ($\mathbf{r}$) and velocity ($\dot{\mathbf{r}}$) of m_2 with respect to m_1 at some time t. The first algorithm presents the sequence in which Kepler's equation (eq. 4.1) and the equations for position and velocity (eqs. 4.15 and 4.16) are solved. The second algorithm, solving Kepler's equation, would normally be embedded in the first algorithm as a subroutine. Note that both algorithms as presented are valid only for elliptical orbits. Refer to Appendix B for the hyperbolic form of Kepler's equation.

4.5.1 Algorithm No. 1: Computing Position and Velocity

Given   The initial conditions for position and velocity $\mathbf{r}_0$ and $\dot{\mathbf{r}}_0$ at an initial time t_0.

Find   The solution for $\mathbf{r}$ and $\dot{\mathbf{r}}$ at time t.

Procedure

▷ Compute the magnitude of $\mathbf{r}_0$ and the energy:

$$r_0 = \|\mathbf{r}_0\|$$
$$h = \frac{1}{2}\dot{\mathbf{r}}_0 \cdot \dot{\mathbf{r}}_0 - \frac{\mu}{r_0}.$$

If $h \geq 0$,   STOP (not elliptical).
If $h < 0$,   CONTINUE.

▷ Compute the angular momentum, its magnitude, associated unit vector $\hat{\mathbf{c}}$, and the semi-latus rectum:

$$\mathbf{c} = \mathbf{r}_0 \times \dot{\mathbf{r}}_0$$
$$c = \|\mathbf{c}\|$$
$$\hat{\mathbf{c}} = \frac{1}{c}\mathbf{c}$$
$$p = \frac{c^2}{\mu}.$$

▷ Compute the Laplace vector, its magnitude, and the unit vectors $\hat{\mathbf{P}}$ and $\hat{\mathbf{Q}}$:

$$\mathbf{P} = -\frac{\mu}{r_0}\mathbf{r}_0 - \mathbf{c} \times \dot{\mathbf{r}}_0$$
$$P = \|\mathbf{P}\|$$

$$\hat{\mathbf{P}} = \frac{1}{P}\,\mathbf{P}$$

$$\hat{\mathbf{Q}} = \hat{\mathbf{c}} \times \hat{\mathbf{P}}.$$

▷ Compute the semi-major axis and the eccentricity,

$$a = -\frac{\mu}{2h}$$

$$e = \frac{P}{\mu}.$$

▷ Compute the time of pericenter passage,

$$e\cos E_0 = 1 - \frac{r_0}{a}$$

$$e\sin E_0 = \frac{\mathbf{r}_0 \cdot \dot{\mathbf{r}}_0}{\sqrt{\mu a}}$$

$$E_0 = \tan^{-1}\left[\frac{e\sin E_0}{e\cos E_0}\right]$$

$$t_\pi = t_0 - \sqrt{\frac{a^3}{\mu}}\,(E_0 - e\sin E_0).$$

▷ Solve Kepler's equation,

$$E - e\sin E = \sqrt{\frac{\mu}{a^3}}\,(t - t_\pi)$$

by iteration for E (see Algorithm No. 2, which follows).

▷ Compute position and velocity using this value of E,

$$\mathbf{r} = a(\cos E - e)\,\hat{\mathbf{P}} + \sqrt{ap}\,\sin E\,\hat{\mathbf{Q}}$$

$$r = \|\mathbf{r}\|$$

$$\dot{\mathbf{r}} = -\frac{\sqrt{\mu a}}{r}\sin E\,\hat{\mathbf{P}} + \frac{\sqrt{\mu p}}{r}\cos E\,\hat{\mathbf{Q}}.$$

End.

4.5.2 Algorithm No. 2: Kepler's Equation—Solution

Given   The current time t, the time of pericenter passage t_π, the semi-major axis a, the eccentricity e, and a first guess for the eccentric anomaly E_1.

Find   The eccentric anomaly E.

Procedure

▷ Compute $F(E_1)$ and $(\frac{dF}{dE})_1$ using the current value for E_1,

$$\left(\frac{dF}{dE}\right)_1 = 1 - e \cos E_1$$

$$F(E_1) = E_1 - e \sin E_1 - n(t - t_\pi).$$

▷ Calculate E,

$$E = E_1 - \frac{F(E_1)}{\left(\frac{dF}{dE}\right)_1}.$$

▷ If $|E - E_1| >$ a desired error tolerance (say, 10^{-8}), then set $E_1 = E$ and go back to the first step.
▷ If E is within our error tolerance, then we have determined E to the desired accuracy.

End.

4.6 Orbital Elements

The Keplerian orbital elements are normally considered to be the following:

- The semi-major axis, a, which was introduced in §3.1.
- The eccentricity, e, also introduced in §3.1.
- The time of pericenter passage, t_π, introduced in §4.1.
- The angle of the ascending node, Ω, from Appendix A.
- The inclination, i, from Appendix A.
- The argument of pericenter, ω, also from Appendix A.

These elements are of course also integrals of motion of the system of differential equations (2.8).

At times we are given the position and velocity at some time and asked to find the integrals (or an equivalent set of orbital elements) for the motion. We state the problem as follows. Given $\mathbf{r}$ and $\dot{\mathbf{r}}$ at time t, find a, e, t_π, i, Ω, and ω. We will point out some of the disadvantages of the Keplerian elements as we proceed and will give a typical example.

a. Compute the energy and semi-major axis:

$$r = \|\mathbf{r}\|$$

$$h = \frac{1}{2}\dot{\mathbf{r}} \cdot \dot{\mathbf{r}} - \frac{\mu}{r}$$

$$a = -\frac{\mu}{2h} \tag{4.20}$$

b. Compute the angular momentum vector, Laplace vector, and the eccentricity:

$$\mathbf{c} = \mathbf{r} \times \dot{\mathbf{r}}$$

$$\mathbf{P} = -\frac{\mu}{r}\mathbf{r} - \mathbf{c} \times \dot{\mathbf{r}}$$

$$P = \|\mathbf{P}\|$$

$$e = \frac{P}{\mu} \tag{4.21}$$

An *alternate* method for calculating e is as follows. Since

$$\frac{1}{\mu}\mathbf{c} \cdot \mathbf{c} = \frac{c^2}{\mu} = a(1 - e^2),$$

we solve for e to arrive at

$$e = \sqrt{1 - \frac{c^2}{\mu a}}. \tag{4.22}$$

c. Compute the time of pericenter passage:

$$e \cos E = 1 - \frac{r}{a}$$

$$e \sin E = \frac{\mathbf{r} \cdot \dot{\mathbf{r}}}{\sqrt{\mu a}}$$

$$E = \tan^{-1}\left(\frac{e \sin E}{e \cos E}\right)$$

$$n = \sqrt{\frac{\mu}{a^3}}$$

$$t_\pi = t - \frac{1}{n}\left(E - e \sin E\right). \tag{4.23}$$

Note that from the above equation we can get another equation for e:

$$e = \sqrt{(e \cos E)^2 + (e \sin E)^2}. \tag{4.24}$$

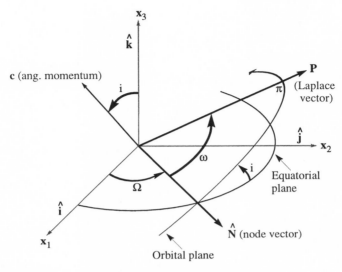

Figure 4.2 Inertial system with orbital elements.

Note that as $e \to 0$ the numerical value for the eccentric anomaly becomes undefined,

$$E \to \tan^{-1}\left(\frac{0}{0}\right),$$

and therefore the time of pericenter passage (t_π) becomes undefined. From §4.1 we know that E is measured from the direction through the pericenter that is undefined for $e = 0$.

d. Compute the inclination and the node angles (fig. 4.2):

$$i = \cos^{-1}(\hat{\mathbf{k}} \cdot \hat{\mathbf{c}}), \tag{4.25}$$

where $\hat{\mathbf{c}} = \mathbf{c}/c$. Note that i can range from $0°$ to $180°$. Orbits with $0° \leq i \leq 90°$ are called *direct orbits* and orbits where $90° \leq i \leq 180°$ are called *retrograde orbits*.

From Appendix A we found that

$$\hat{\mathbf{i}} \cdot \hat{\mathbf{c}} = \sin \Omega \sin i$$

and

$$\hat{\mathbf{j}} \cdot \hat{\mathbf{c}} = -\cos \Omega \sin i.$$

Divide the first equation by the second to get

$$\tan \Omega = \frac{\sin \Omega \sin i}{\cos \Omega \sin i} = \frac{\hat{\mathbf{i}} \cdot \hat{\mathbf{c}}}{-\hat{\mathbf{j}} \cdot \hat{\mathbf{c}}},$$

which results in

$$\Omega = \tan^{-1} \left(\frac{\hat{\mathbf{i}} \cdot \hat{\mathbf{c}}}{-\hat{\mathbf{j}} \cdot \hat{\mathbf{c}}} \right). \tag{4.26}$$

As can be seen from figure 4.2, Ω can range from $0°$ to $360°$. Note the numerical problem that exists as $i \to 0$, $\sin i \to 0$, and therefore from equation (4.26),

$$\Omega \to \tan^{-1} \left(\frac{0}{0} \right),$$

and Ω becomes *undefined*. The problem here is quite basic, as the orbital plane approaches the equatorial plane, they intersect everywhere in the equatorial plane.

e. Compute the argument of the pericenter:

$$\cos \omega = \hat{\mathbf{N}} \cdot \hat{\mathbf{P}}$$

$$\sin \omega = \hat{\mathbf{c}} \cdot \hat{\mathbf{N}} \times \hat{\mathbf{P}}$$

$$\omega = \tan^{-1} \left(\frac{\sin \omega}{\cos \omega} \right). \tag{4.27}$$

From figure 4.2 we see that ω ranges from $0°$ to $360°$. Note the numerical problem that as $e \to 0$, $\hat{\mathbf{P}}$ becomes undefined since

$$\hat{\mathbf{P}} = \frac{1}{\|\mathbf{P}\|} \mathbf{P} = \frac{1}{(\mu e)} \mathbf{P}.$$

Therefore ω also becomes undefined. Also, as $i \to 0$, $\hat{\mathbf{N}}$ becomes undefined and so ω becomes undefined.

We illustrate with a numerical example from Stiefel and Scheifele[64].

Given   The position and velocity vectors,

$$\mathbf{r} = \begin{pmatrix} x_1 \\ x_2 \\ x_3 \end{pmatrix} = \begin{pmatrix} 0.0 \text{ km} \\ -5888.9727 \text{ km} \\ -3400.0 \text{ km} \end{pmatrix}$$

$$\dot{\mathbf{r}} = \begin{pmatrix} \dot{x}_1 \\ \dot{x}_2 \\ \dot{x}_3 \end{pmatrix} = \begin{pmatrix} 10.691338 \text{ km/sec} \\ 0.0 \text{ km/sec} \\ 0.0 \text{ km/sec} \end{pmatrix}$$

at time t. Let $\mu = 398601.0 \text{ km}^3/\text{sec}^2$.

Find   The orbital elements a, e, t_π, i, Ω and ω.

Procedure

▷ Calculate the radius r,

$$r = \|\mathbf{r}\| = \sqrt{x_1^2 + x_2^2 + x_3^2} = 6800 \text{ km}.$$

▷ The energy (h) is given by

$$h = \frac{1}{2}\dot{\mathbf{r}} \cdot \dot{\mathbf{r}} - \frac{\mu}{r} = -1.4654400 \text{ km}^2/\text{sec}^2.$$

▷ The semi-major axis is

$$a = -\frac{\mu}{2h} = 136000.45 \text{ km}.$$

▷ Compute the angular momentum,

$$\mathbf{c} = \mathbf{r} \times \dot{\mathbf{r}} = \begin{vmatrix} \hat{\mathbf{i}} & \hat{\mathbf{j}} & \hat{\mathbf{k}} \\ 0 & x_2 & x_3 \\ \dot{x}_1 & 0 & 0 \end{vmatrix}$$

$$= (-36350.55 \text{ km}^2/\text{sec})\,\hat{\mathbf{j}} + (62961.0 \text{ km}^2/\text{sec})\,\hat{\mathbf{k}},$$

and the magnitude,

$$c = \|\mathbf{c}\| = 7.2701119 \times 10^4 \text{ km}^2/\text{sec}.$$

▷ Now compute the Laplace vector,

$$\mathbf{P} = -\frac{\mu}{r}\mathbf{r} - \mathbf{c} \times \dot{\mathbf{r}}$$

$$= (-327938.74 \text{ km}^3/\text{sec}^2)\,\hat{\mathbf{j}} - (189335.52 \text{ km}^3/\text{sec}^2)\,\hat{\mathbf{k}},$$

and the magnitude,

$$P = \|\mathbf{P}\| = 3.7867104 \times 10^5 \text{ km}^3/\text{sec}^2.$$

▷ Also, the unit vector $\hat{\mathbf{P}}$ is

$$\hat{\mathbf{P}} = \frac{1}{P}\mathbf{P} = (-0.8660254)\,\hat{\mathbf{j}} - (0.5)\,\hat{\mathbf{k}}.$$

▷ The eccentricity is

$$e = \frac{P}{\mu} = 0.95.$$

Compute e by an alternate method,

$$e = \sqrt{1 - \frac{c^2}{\mu a}} = 0.95.$$

▷ Calculate the eccentric anomaly,

$$e \cos E = 1 - \frac{r}{a} = 0.95$$

$$e \sin E = \frac{\mathbf{r} \cdot \dot{\mathbf{r}}}{\sqrt{\mu a}} = 0$$

$$\tan E = \frac{e \sin E}{e \cos E} = \frac{0}{0.95} = 0$$

$$E = \tan^{-1}(\tan E) = 0.$$

▷ From Kepler's equation,

$$\sqrt{\frac{\mu}{a^3}} \, (t - t_\pi) = E - e \sin E = 0$$

$$t - t_\pi = 0$$

$$t_\pi = t.$$

Again compute e by an alternate method,

$$e^2 = (e \cos E)^2 + (e \sin E)^2 = (0.95)^2$$

$$e = 0.95.$$

▷ We calculate i from $\cos i = \hat{\mathbf{k}} \cdot \hat{\mathbf{c}}$,

$$\hat{\mathbf{c}} = \frac{\mathbf{c}}{c} = \frac{-36350.55 \, \hat{\mathbf{j}} + 62961.0 \, \hat{\mathbf{k}}}{72701.119}$$

$$= (-0.5) \hat{\mathbf{j}} + (0.866025) \hat{\mathbf{k}}$$

$$\cos i = \hat{\mathbf{k}} \cdot \hat{\mathbf{c}} = 0.866025$$

$$i = \cos^{-1}(0.866025) = 30°.$$

▷ Calculate the angle of the ascending node (Ω),

$$\hat{\mathbf{i}} \cdot \hat{\mathbf{c}} = 0$$

$$\hat{\mathbf{j}} \cdot \hat{\mathbf{c}} = -0.5$$

$$\tan \Omega = \frac{\sin \Omega \sin i}{\cos \Omega \sin i} = \frac{\hat{\mathbf{i}} \cdot \hat{\mathbf{c}}}{-\hat{\mathbf{j}} \cdot \hat{\mathbf{c}}} = \frac{0}{0.5}$$

$$\Omega = 0°.$$

▷ Finally calculate the argument of perigee (ω),

$$\hat{\mathbf{N}} = \cos \Omega \, \hat{\mathbf{i}} + \sin \Omega \, \hat{\mathbf{j}} = \hat{\mathbf{i}}$$

$$\cos \omega = \hat{\mathbf{N}} \cdot \hat{\mathbf{P}} = \hat{\mathbf{i}} \cdot (P_2 \hat{\mathbf{j}} + P_3 \hat{\mathbf{k}}) = 0$$

$$\sin \omega = \hat{\mathbf{c}} \cdot \hat{\mathbf{N}} \times \hat{\mathbf{P}}.$$

Therefore,

$$\sin \omega = -c_2 P_3 + c_3 P_2$$

$$\sin \omega = -1$$

$$\omega = \tan^{-1} \left(\frac{\sin \omega}{\cos \omega} \right) = \tan^{-1} \left(\frac{-1}{0} \right)$$

$$\omega = 270°.$$

End.

In reference [64] Stiefel and Scheifele define an element to be "any quantity which during a pure Kepler motion is a linear function of the independent variable." This definition includes the integrals, or constants, of the motion. For example, we can write the semi-major axis, which is an integral of the motion, as

$$a = a_0 + \dot{a} \, t;$$

but since $\dot{a} = 0$, we have

$$a = a_0 = constant.$$

In most instances we can regard the terms "element" and "integral" to be synonymous. But note that the mean anomaly M introduced in §4.1 is also an element, since

$$M = n(t - t_\pi),$$

with the mean motion n defined as

$$n = \sqrt{\frac{\mu}{a^3}} = constant,$$

is certainly a linear function of time.

4.7 Other Orbital Element Systems

The derivation of the transformation from Keplerian elements to Delaunay elements and then to Poincaré elements is in the realm of Hamiltonian mechanics. These transformations will be given without derivation, since an introduction to Hamiltonian mechanics is beyond the scope of this book.

4.7.1 Delaunay Elements

The Delaunay elements are due to Charles-Eugene Delaunay, who was born on 9 April 1816 in Lusiguy, France. He died at sea near Cherbourg on 5 August 1872. He developed a solution for the motion of the Moon. Using this solution, he noted a discrepancy between computation and observations, which led him, in 1865, to postulate to that tidal friction due to the orbital motion of the moon was causing the rotational rate of the Earth to decrease. This theory was later validated. The transformation from Keplerian to Delaunay elements is

$$l_D = M = \sqrt{\frac{\mu}{a^3}}\,(t - t_\pi)$$
$$g_D = \omega$$
$$h_D = \Omega$$
$$L_D = \sqrt{\mu a}$$
$$G_D = \sqrt{\mu a(1 - e^2)} = c$$
$$H_D = \sqrt{\mu a(1 - e^2)}\,\cos i.$$

This transformation is just a change in notation of the Keplerian elements for the mean anomaly, argument of perigee, ascending node, and angular momentum. The subscript (D) on the Delaunay elements is optional and not usually needed. However, we use the subscript here to avoid confusion with previously defined elements.

4.7.2 Poincaré Elements

The Poincaré elements are due to Jules Henri Poincaré, who was born in Nancy, France, on 29 April 1854. He died in Paris on 17 July 1912. Poincaré is known for his rigorous definition and analysis in celestial mechanics. He published about one hundred papers on the subject, most of which are contained in his three-volume work, *Les Methodes nouvelles de la mécanique céleste.* The transformation from Delaunay to Poincaré elements is

$$\rho_1 = L_D$$
$$\rho_2 = \sqrt{2(L_D - G_D)} \cos(g_D + h_D)$$
$$\rho_3 = \sqrt{2(G_D - H_D)} \cos h_D$$
$$\sigma_1 = l_D + g_D + h_D$$
$$\sigma_2 = -\sqrt{2(L_D - G_D)} \sin(g_D + h_D)$$
$$\sigma_3 = -\sqrt{2(G_D - H_D)} \sin h_D.$$

The advantage of the Poincaré elements is that, in contrast to the Keplerian elements that were discussed in §4.6, they remain defined for small inclination and eccentricity values. Consider the following cases:

Case (a)—Small inclination. As $i \rightarrow 0$, $H_D \rightarrow G_D$, and therefore ρ_3 and $\sigma_3 \rightarrow 0$. Even though the node angle ($h_D = \Omega$) becomes undefined it will not matter since the sum $g_D + h_D$ is used in the elements σ_1, σ_2 and ρ_2.

Case (b)—Small eccentricity. As $e \rightarrow 0$, $G_D \rightarrow L_D$, and therefore ρ_2 and $\sigma_2 \rightarrow 0$. Even though the argument of perigee ($g_D = \omega$) becomes undefined for this case it will not matter since g_D appears only in the sum $g_D + h_D$ in the element σ_1.

5

THE f AND g FUNCTIONS

In this chapter we will develop solutions of the two-body problem expressed in inertial cartesian coordinates. In principle, the reduction of the two-body problem to six independent integrals of the motion is sufficient for a solution. However, in many practical applications it is desirable to have the solution in inertial cartesian coordinates. We will show that the solution given in chapter 4 in terms of the unit vectors $\hat{\mathbf{P}}$, $\hat{\mathbf{Q}}$, $\hat{\mathbf{c}}$ can be transformed into a solution of the form

$$\mathbf{r} = \mathbf{r}_0 \, f + \dot{\mathbf{r}}_0 \, g$$

$$\dot{\mathbf{r}} = \mathbf{r}_0 \, \dot{f} + \dot{\mathbf{r}}_0 \, \dot{g},$$

where f, g, $\dot{f}$, and $\dot{g}$ are functions of the initial conditions and the time; that is,

$$f = f(\mathbf{r}_0, \dot{\mathbf{r}}_0, t)$$

and so forth. The solution in this form places minimum emphasis on the orbital elements—Keplerian elements, for example. In fact, when the solution is expressed in the f and g functions, the initial conditions $(\mathbf{r}_0, \dot{\mathbf{r}}_0)$ meet all of the requirements of orbital elements. This is exactly what we should expect since we showed in chapter 4 that the Keplerian elements could be obtained from the position and velocity.

We should also comment here that although f, g, $\dot{f}$, and $\dot{g}$ are functionally expressed in time, in practice they are expressed most often as implicit functions of time. For example,

$$f = f(\mathbf{r}_0, \dot{\mathbf{r}}_0, t(\alpha))$$

and so forth, where α is one of the anomaly angles or some other auxiliary variable. These expressions of implicit dependence upon time simply reflect the basic transcendental relationship of the two-body problem to time.

5.1 f and g Functions—Development

This section follows that of Battin [4]. We showed in chapter 4 that the position and velocity could be expressed in the orbital plane system defined by the unit vectors $\hat{\mathbf{P}}$, $\hat{\mathbf{c}}$, and $\hat{\mathbf{Q}}$. That is,

$$\mathbf{r} = l\,\hat{\mathbf{P}} + m\,\hat{\mathbf{Q}} \tag{5.1}$$

$$\dot{\mathbf{r}} = \dot{l}\,\hat{\mathbf{P}} + \dot{m}\,\hat{\mathbf{Q}}. \tag{5.2}$$

The functions l, m, $\dot{l}$, and $\dot{m}$ are clearly identifiable in equations (4.6) and (4.9) in terms of the true anomaly,

$$l = r\cos\phi \tag{5.3}$$

$$m = r\sin\phi \tag{5.4}$$

$$\dot{l} = -\sqrt{\frac{\mu}{p}}\,\sin\phi \tag{5.5}$$

$$\dot{m} = \sqrt{\frac{\mu}{p}}\,(e + \cos\phi). \tag{5.6}$$

Another set of these functions as shown in equations (4.15) and (4.16) in terms of the eccentric anomaly is

$$l = a(\cos E - e) \tag{5.7}$$

$$m = \sqrt{ap}\,\sin E \tag{5.8}$$

$$\dot{l} = -\frac{\sqrt{\mu a}}{r}\sin E \tag{5.9}$$

$$\dot{m} = \frac{\sqrt{\mu p}}{r}\cos E. \tag{5.10}$$

Many other sets for l, m, $\dot{l}$, and $\dot{m}$ exist in the literature.

The disadvantage in the use of equations (5.1) and (5.2) and either (5.3)–(5.6) or (5.7)–(5.10) in the solution for position and velocity is that the integrals must be computed from the initial conditions $\mathbf{r}_0$, $\dot{\mathbf{r}}_0$, and t_0 before the solution can be found. The algorithm we presented in chapter 4 for computing Keplerian elements had numerical problems when e and i were small.

We now show that equations (5.1) and (5.2) for $\mathbf{r}$ and $\dot{\mathbf{r}}$ can be transformed into a solution in the initial conditions $\mathbf{r}_0$, $\dot{\mathbf{r}}_0$, and t_0. At the initial time t_0, equations (5.1) and (5.2) become

$$\mathbf{r}_0 = l_0 \,\hat{\mathbf{P}} + m_0 \,\hat{\mathbf{Q}} \tag{5.11}$$

$$\dot{\mathbf{r}}_0 = \dot{l}_0 \,\hat{\mathbf{P}} + \dot{m}_0 \,\hat{\mathbf{Q}}. \tag{5.12}$$

Now solve equations (5.11) and (5.12) for $\hat{\mathbf{P}}$ and $\hat{\mathbf{Q}}$. This can be done quite easily in component form where the vectors are assumed to be in the inertial coordinate system centered at one of the masses (see fig. 4.2). Let

$$\mathbf{r}_0 = \hat{\mathbf{i}}\, x_0 + \hat{\mathbf{j}}\, y_0 + \hat{\mathbf{k}}\, z_0$$

$$\dot{\mathbf{r}}_0 = \hat{\mathbf{i}}\, \dot{x}_0 + \hat{\mathbf{j}}\, \dot{y}_0 + \hat{\mathbf{k}}\, \dot{z}_0$$

$$\hat{\mathbf{P}} = \hat{\mathbf{i}}\, P_x + \hat{\mathbf{j}}\, P_y + \hat{\mathbf{k}}\, P_z$$

$$\hat{\mathbf{Q}} = \hat{\mathbf{i}}\, Q_x + \hat{\mathbf{j}}\, Q_y + \hat{\mathbf{k}}\, Q_z,$$

where

$$P_x = \hat{\mathbf{i}} \cdot \hat{\mathbf{P}}, \text{ etc.}$$

$$Q_x = \hat{\mathbf{i}} \cdot \hat{\mathbf{Q}}, \text{ etc.}$$

For the x component, equations (5.11) and (5.12) become

$$x_0 = l_0 \, P_x + m_0 \, Q_x$$

$$\dot{x}_0 = \dot{l}_0 \, P_x + \dot{m}_0 \, Q_x,$$

or in matrix notation,

$$\begin{pmatrix} x_0 \\ \dot{x}_0 \end{pmatrix} = \begin{pmatrix} l_0 & m_0 \\ \dot{l}_0 & \dot{m}_0 \end{pmatrix} \begin{pmatrix} P_x \\ Q_x \end{pmatrix}. \tag{5.13}$$

Now invert equation (5.13) and solve for P_x and Q_x:

$$\begin{pmatrix} P_x \\ Q_x \end{pmatrix} = \frac{1}{\Delta} \begin{pmatrix} \dot{m}_0 & -m_0 \\ -\dot{l}_0 & l_0 \end{pmatrix} \begin{pmatrix} x_0 \\ \dot{x}_0 \end{pmatrix}$$

or

$$P_x = \frac{1}{\Delta} (\dot{m}_0 x_0 - m_0 \dot{x}_0) \tag{5.14}$$

$$Q_x = \frac{1}{\Delta} (-\dot{l}_0 x_0 + l_0 \dot{x}_0), \tag{5.15}$$

where the determinant of the matrix is

$$\Delta = l_0 \dot{m}_0 - m_0 \dot{l}_0.$$

The derivation for the y and z components follows the same procedure. Therefore, we can generalize (5.14) and (5.15) to

$$\hat{\mathbf{P}} = \frac{1}{\Delta}(\dot{m}_0 \, \mathbf{r}_0 - m_0 \, \dot{\mathbf{r}}_0) \qquad (5.16)$$

$$\hat{\mathbf{Q}} = \frac{1}{\Delta}(-\dot{l}_0 \, \mathbf{r}_0 + l_0 \, \dot{\mathbf{r}}_0). \qquad (5.17)$$

To determine Δ we use the angular momentum. From equation (2.14),

$$\mathbf{c} = \mathbf{r}_0 \times \dot{\mathbf{r}}_0.$$

By using equations (5.11) and (5.12), and $\hat{\mathbf{P}} \times \hat{\mathbf{P}} = 0$ and $\hat{\mathbf{Q}} \times \hat{\mathbf{Q}} = 0$, we get

$$\mathbf{c} = l_0 \dot{m}_0 (\hat{\mathbf{P}} \times \hat{\mathbf{Q}}) + m_0 \dot{l}_0 (\hat{\mathbf{Q}} \times \hat{\mathbf{P}}).$$

Since $\hat{\mathbf{c}} = \hat{\mathbf{P}} \times \hat{\mathbf{Q}}$, we have

$$\mathbf{c} = (l_0 \dot{m}_0 - m_0 \dot{l}_0)\hat{\mathbf{c}} = c\,\hat{\mathbf{c}}. \qquad (5.18)$$

Thus we have shown that

$$c = \Delta = l_0 \dot{m}_0 - m_0 \dot{l}_0 = \sqrt{\mu p}. \qquad (5.19)$$

Equations (5.16) and (5.17) then become

$$\hat{\mathbf{P}} = \frac{1}{\sqrt{\mu p}}(\dot{m}_0 \, \mathbf{r}_0 - m_0 \, \dot{\mathbf{r}}_0) \qquad (5.20)$$

$$\hat{\mathbf{Q}} = \frac{1}{\sqrt{\mu p}}(-\dot{l}_0 \, \mathbf{r}_0 + l_0 \, \dot{\mathbf{r}}_0). \qquad (5.21)$$

Now substitute equations (5.20) and (5.21) back into equations (5.1) and (5.2), eliminating $\hat{\mathbf{P}}$ and $\hat{\mathbf{Q}}$. We do this first for the position vector $\mathbf{r}$,

$$\mathbf{r} = \frac{1}{\sqrt{\mu p}}[l(\dot{m}_0 \, \mathbf{r}_0 - m_0 \, \dot{\mathbf{r}}_0) + m(-\dot{l}_0 \, \mathbf{r}_0 + l_0 \, \dot{\mathbf{r}}_0)].$$

Collect coefficients of $\mathbf{r}_0$ and $\dot{\mathbf{r}}_0$ and obtain

$$\mathbf{r} = \frac{1}{\sqrt{\mu p}}(l\dot{m}_0 - m\dot{l}_0)\mathbf{r}_0 + \frac{1}{\sqrt{\mu p}}(-lm_0 + ml_0)\dot{\mathbf{r}}_0. \qquad (5.22)$$

We could get $\dot{\mathbf{r}}$ in a similar manner by substituting $\hat{\mathbf{P}}$ and $\hat{\mathbf{Q}}$ into the $\dot{\mathbf{r}}$ equation (eq. 5.2). But it is easier to take the time derivative of equation (5.22) to obtain

$$\dot{\mathbf{r}} = \frac{1}{\sqrt{\mu p}}(\dot{l}\dot{m}_0 - \dot{m}\dot{l}_0)\mathbf{r}_0 + \frac{1}{\sqrt{\mu p}}(-\dot{l}m_0 + \dot{m}l_0)\dot{\mathbf{r}}_0. \tag{5.23}$$

Now we introduce the functions

$$f = \frac{1}{\sqrt{\mu p}}(l\dot{m}_0 - m\dot{l}_0) \tag{5.24}$$

$$g = \frac{1}{\sqrt{\mu p}}(-lm_0 + ml_0) \tag{5.25}$$

and their derivatives

$$\dot{f} = \frac{1}{\sqrt{\mu p}}(\dot{l}\dot{m}_0 - \dot{m}\dot{l}_0) \tag{5.26}$$

$$\dot{g} = \frac{1}{\sqrt{\mu p}}(-\dot{l}m_0 + \dot{m}l_0). \tag{5.27}$$

Using equations (5.24)–(5.27), equations (5.22) and (5.23) become

$$\mathbf{r} = f\,\mathbf{r}_0 + g\,\dot{\mathbf{r}}_0 \tag{5.28}$$

$$\dot{\mathbf{r}} = \dot{f}\,\mathbf{r}_0 + \dot{g}\,\dot{\mathbf{r}}_0. \tag{5.29}$$

The form of $f, g, \dot{f}$, and $\dot{g}$ will depend upon the form of the equations for $l, m, \dot{l}$, and $\dot{m}$ which are used. In §5.2 and §5.3 we derive $f, g, \dot{f}$, and $\dot{g}$ in terms of true anomaly and eccentric anomaly.

From equations (5.28) and (5.29) we can derive a relation among the functions $f, g, \dot{f}$, and $\dot{g}$. From the angular momentum,

$$\mathbf{c} = \mathbf{r} \times \dot{\mathbf{r}}.$$

Using $\mathbf{r}_0 \times \mathbf{r}_0 = 0$, and $\dot{\mathbf{r}}_0 \times \dot{\mathbf{r}}_0 = 0$, we obtain

$$\mathbf{c} = f\dot{g}(\mathbf{r}_0 \times \dot{\mathbf{r}}_0) + g\dot{f}(\dot{\mathbf{r}}_0 \times \mathbf{r}_0).$$

Since $\mathbf{c} = \mathbf{r} \times \dot{\mathbf{r}} = \mathbf{r}_0 \times \dot{\mathbf{r}}_0 = constant$,

$$\mathbf{c} = (f\dot{g} - g\dot{f})\mathbf{c}. \tag{5.30}$$

Equation (5.30) then requires that

$$f\dot{g} - g\dot{f} = 1. \tag{5.31}$$

Therefore, given any three of the four functions f, g, $\dot{f}$, or $\dot{g}$, we can solve for the remaining one.

5.2 f and g Functions—True Anomaly

We now substitute the functions l, m, $\dot{l}$, and $\dot{m}$ given in equations (5.3)–(5.6) into equations (5.24)–(5.27) for f, g, $\dot{f}$, and $\dot{g}$. Equation (5.24) becomes

$$f = \frac{1}{\sqrt{\mu p}}(l\dot{m}_0 - m\dot{l}_0)$$

$$= \frac{1}{\sqrt{\mu p}} \left[\overbrace{(r\cos\phi)}^{l} \overbrace{\sqrt{\frac{\mu}{p}}(e + \cos\phi_0)}^{\dot{m}_0} - \underbrace{(r\sin\phi)}_{m} \underbrace{\left(-\sqrt{\frac{\mu}{p}}\sin\phi_0\right)}_{\dot{l}_0} \right]$$

$$f = \frac{r}{\sqrt{\mu p}}\sqrt{\frac{\mu}{p}} [\cos\phi(e + \cos\phi_0) + \sin\phi\sin\phi_0]$$

$$f = \frac{r}{p} [e\cos\phi + \cos\phi\cos\phi_0 + \sin\phi\sin\phi_0]$$

We have already shown that

$$r = \frac{p}{1 + e\cos\phi};$$

therefore

$$e\cos\phi = \frac{p}{r} - 1.$$

We also employ the trigonometric identity

$$\cos(\phi - \phi_0) = \cos\phi\cos\phi_0 + \sin\phi\sin\phi_0.$$

Substitute these two equations into the equation for f:

$$f = \frac{r}{p}\left[\frac{p}{r} - 1 + \cos(\phi - \phi_0)\right].$$

Now define

$$\Delta\phi = \phi - \phi_0, \tag{5.32}$$

and the equation for f becomes

$$f = 1 - \frac{r}{p}(1 - \cos \Delta\phi).$$ (5.33)

For g, substitute equations (5.3) and (5.4) into equation (5.25):

$$g = \frac{1}{\sqrt{\mu p}}(-l\, m_0 + m\, l_0)$$

$$= \frac{1}{\sqrt{\mu p}}\left[\underbrace{-r \cos \phi}_{l}\ \underbrace{r_0 \sin \phi_0}_{m_0} + \underbrace{r \sin \phi}_{m}\ \underbrace{r_0 \cos \phi_0}_{l_0} \right]$$

$$= \frac{r r_0}{\sqrt{\mu p}}(\sin \phi \cos \phi_0 - \cos \phi \sin \phi_0).$$

Use the trigonometric identity

$$\sin(\phi - \phi_0) = \sin \phi \cos \phi_0 - \cos \phi \sin \phi_0$$

to arrive at

$$g = \frac{r r_0}{\sqrt{\mu p}} \sin \Delta\phi.$$ (5.34)

We derive the equation for $\dot{g}$ by substituting equations (5.3)–(5.6) into equation (5.27):

$$\dot{g} = \frac{1}{\sqrt{\mu p}}(-\dot{l} m_0 + \dot{m} l_0)$$

$$= \frac{1}{\sqrt{\mu p}}\left[-\overbrace{\left(-\sqrt{\frac{\mu}{p}} \sin \phi\right)}^{\dot{l}}\ \overbrace{r_0 \sin \phi_0}^{m} + \underbrace{\sqrt{\frac{\mu}{p}}(e + \cos \phi)}_{\dot{m}}\ \underbrace{r_0 \cos \phi_0}_{l_0} \right]$$

$$= \frac{r_0}{\sqrt{\mu p}}\sqrt{\frac{\mu}{p}}\ [\sin \phi \sin \phi_0 + \cos \phi \cos \phi_0 + e \cos \phi_0]$$

$$= \frac{r_0}{p}\ [\cos(\phi - \phi_0) + e \cos \phi_0].$$

As we did for f, we use

$$e \cos \phi_0 = \frac{p}{r_0} - 1$$

and the definition for $\Delta\phi$ to arrive at

$$\dot{g} = 1 - \frac{r_0}{p}(1 - \cos\Delta\phi). \tag{5.35}$$

Note the similarity between equations (5.33) for f and (5.35) for $\dot{g}$.

To get an expression for $\dot{f}$ we substitute equations (5.33)–(5.35) into equation (5.31):

$$-g\dot{f} = 1 - f\dot{g}$$

$$-g\dot{f} = 1 - \left[1 - \frac{r}{p}(1 - \cos\Delta\phi)\right]\left[1 - \frac{r_0}{p}(1 - \cos\Delta\phi)\right].$$

After further reduction,

$$-g\dot{f} = \frac{1 - \cos\Delta\phi}{p}\left[r + r_0 - \frac{rr_0}{p}(1 - \cos\Delta\phi)\right].$$

Dividing by g we obtain

$$-\dot{f} = \frac{\sqrt{\mu p}}{rr_0 \sin\Delta\phi}\frac{1 - \cos\Delta\phi}{p}\left[r + r_0 - \frac{rr_0}{p}(1 - \cos\Delta\phi)\right]$$

$$\dot{f} = \sqrt{\frac{\mu}{p}}\left(\frac{1 - \cos\Delta\phi}{\sin\Delta\phi}\right)\left[\frac{1}{p}(1 - \cos\Delta\phi) - \frac{1}{r_0} - \frac{1}{r}\right]. \tag{5.36}$$

Summarizing, the f and g functions and their derivatives in the true anomaly are

$$f = 1 - \frac{r}{p}(1 - \cos\Delta\phi) \tag{5.37}$$

$$g = \frac{rr_0}{\sqrt{\mu p}}\sin\Delta\phi \tag{5.38}$$

$$\dot{g} = 1 - \frac{r_0}{p}(1 - \cos\Delta\phi) \tag{5.39}$$

$$\dot{f} = \sqrt{\frac{\mu}{p}}\left(\frac{1 - \cos\Delta\phi}{\sin\Delta\phi}\right)\left[\frac{1}{p}(1 - \cos\Delta\phi) - \frac{1}{r_0} - \frac{1}{r}\right]. \tag{5.40}$$

Since there is no convenient relation between time and $\Delta\phi$, this form is not normally used in the solution of the initial value problem. It is, however, useful in the solution of two-point boundary value problems. We will use the $\Delta\phi$ f and g functions in chapter 6 when we formulate the solution to Lambert's problem.

5.3 f and g Functions—Eccentric Anomaly

We now substitute the functions $l, m, \dot{l}$, and $\dot{m}$ given in equations (5.7)–(5.10) into equation (5.24) for f,

$$f = \frac{1}{\sqrt{\mu p}}(l\dot{m}_0 - m\dot{l}_0)$$

$$= \frac{1}{\sqrt{\mu p}}\left[\overbrace{a(\cos E - e)}^{l}\;\overbrace{\frac{\sqrt{\mu p}}{r_0}\cos E_0}^{\dot{m}_0} - \underbrace{\sqrt{ap}\sin E}_{m}\underbrace{\left(-\frac{\sqrt{\mu a}}{r_0}\sin E_0\right)}_{\dot{l}_0}\right]$$

$$= \frac{a}{r_0}[(\cos E - e)\cos E_0 + \sin E \sin E_0]$$

$$f = \frac{a}{r_0}[-e\cos E_0 + \cos E \cos E_0 + \sin E \sin E_0].$$

Recall that $r_0 = a(1 - e\cos E_0)$, so

$$e\cos E_0 = 1 - \frac{r_0}{a}.$$

The equation for f then becomes

$$f = \frac{a}{r_0}\left[\frac{r_0}{a} - 1 + \cos(E - E_0)\right].$$

Defining

$$\Delta E = E - E_0, \tag{5.41}$$

the equation for f becomes finally

$$f = 1 - \frac{a}{r_0}(1 - \cos(\Delta E)). \tag{5.42}$$

Note that we can get $\dot{f}$ by taking the derivative of equation (5.42),

$$\dot{f} = \frac{a}{r_0}\frac{d}{dt}(\cos \Delta E)$$

$$= -\frac{a}{r_0}\sin \Delta E \frac{dE}{dt}.$$

From equation (4.12) we have

$$\frac{dE}{dt} = \frac{\sqrt{\mu/a}}{r},$$

so the equation for $\dot{f}$ becomes

$$\dot{f} = -\frac{\sqrt{\mu a}}{r r_0} \sin \Delta E. \tag{5.43}$$

Now substitute equations (5.7)–(5.10) into equation (5.27) for $\dot{g}$:

$$\dot{g} = \frac{1}{\sqrt{\mu p}}(-\dot{l}m_0 + \dot{m}l_0)$$

$$= \frac{1}{\sqrt{\mu p}}\left[\overbrace{\frac{\sqrt{\mu a}}{r}\sin E}^{-\dot{l}}\overbrace{\sqrt{ap}\sin E_0}^{m} + \frac{\sqrt{\mu p}}{r}\cos E \underbrace{a(\cos E_0 - e)}_{l_0}\right]$$
$$\hspace{6cm}\underbrace{\hspace{3cm}}_{\dot{m}}$$

$$= \frac{1}{\sqrt{\mu p}}\frac{a\sqrt{\mu p}}{r}\left[\sin E \sin E_0 + \cos E \cos E_0 - \underbrace{e \cos E}_{1-(r/a)}\right]$$

$$\dot{g} = \frac{a}{r}\left[\cos(E - E_0) + \frac{r}{a} - 1\right].$$

Finally,

$$\dot{g} = 1 - \frac{a}{r}(1 - \cos \Delta E). \tag{5.44}$$

To obtain an expression for g, we integrate equation (5.44). Multiply (5.44) by dt/dE,

$$\frac{dg}{dt}\frac{dt}{dE} = \left[1 - \frac{a}{r}(1 - \cos \Delta E)\right]\frac{dt}{dE}.$$

From equation (4.12),

$$\frac{dt}{dE} = \sqrt{\frac{a}{\mu}}\, r.$$

Therefore,

$$\frac{dg}{dE} = \frac{dt}{dE} - \frac{a}{r}(1 - \cos \Delta E)\sqrt{\frac{a}{\mu}}r$$

$$dg = dt - \sqrt{\frac{a^3}{\mu}}\,(1 - \cos \Delta E)dE.$$

Observe that $dE = d(E - E_0) = d(\Delta E)$, so we can integrate the above equation directly to get

$$g = t - \sqrt{\frac{a^3}{\mu}} \, (\Delta E - \sin \Delta E) + constant. \qquad (5.45)$$

In order to evaluate the constant we need the initial value of g. Recall that

$$\mathbf{r} = f \, \mathbf{r}_0 + g \, \dot{\mathbf{r}}_0.$$

When $t = t_0$, $\mathbf{r} = \mathbf{r}_0$, therefore $f(t_0) = 1$ and $g(t_0) = 0$. Also when $t = t_0$, $E = E_0$. We evaluate equation (5.45) at t_0,

$$g(t_0) = 0 = t_0 + constant$$

or

$$constant = -t_0.$$

Equation (5.45) then becomes

$$g = t - t_0 - \sqrt{\frac{a^3}{\mu}} \, (\Delta E - \sin \Delta E). \qquad (5.46)$$

Summarizing, the f and g functions and their derivatives in the eccentric anomaly are

$$f = 1 - \frac{a}{r_0} \, (1 - \cos(\Delta E))$$

$$\dot{f} = -\frac{\sqrt{\mu a}}{r r_0} \sin \Delta E$$

$$g = t - t_0 - \sqrt{\frac{a^3}{\mu}} \, (\Delta E - \sin \Delta E)$$

$$\dot{g} = 1 - \frac{a}{r} (1 - \cos \Delta E).$$

Equation (5.46) indicates that we also need to get Kepler's equation in a form that relates t to ΔE. From equation (4.12),

$$\frac{dt}{dE} = \sqrt{\frac{a}{\mu}} \, r,$$

and since $r = a(1 - e \cos E)$,

$$dt = \sqrt{\frac{a^3}{\mu}} \, (1 - e \cos E) dE,$$

which we integrate,

$$\int_{t_0}^{t} d\tau = \sqrt{\frac{a^3}{\mu}} \int_{E_0}^{E} (1 - e \cos \varepsilon) \, d\varepsilon,$$

where τ and ε are substitute variables of integration for t and E.

$$\tau \Big|_{t_0}^{t} = \sqrt{\frac{a^3}{\mu}} \left(\varepsilon - e \sin \varepsilon \right) \Big|_{E_0}^{E}$$

$$t - t_0 = \sqrt{\frac{a^3}{\mu}} \, [\overbrace{E - E_0}^{\Delta E} - e(\sin E - \sin E_0)]. \tag{5.47}$$

We want to eliminate $\sin E$ since we want the formulation to be in ΔE. Using the elementary trigonometric identity,

$$\sin E = \cos E_0 \sin \overbrace{(E - E_0)}^{\Delta E} + \sin E_0 \cos \overbrace{(E - E_0)}^{\Delta E}.$$

So the term in equation (5.47),

$$e(\sin E - \sin E_0) = e \, [\cos E_0 \sin \Delta E + \sin E_0 \cos \Delta E - \sin E_0]$$

$$= (e \cos E_0) \sin \Delta E + (e \sin E_0)(\cos \Delta E - 1).$$

Recall that

$$e \cos E_0 = 1 - (r_0/a)$$

$$e \sin E_0 = \frac{\mathbf{r}_0 \cdot \dot{\mathbf{r}}_0}{\sqrt{\mu a}}.$$

Using these, the above equation becomes

$$e(\sin E - \sin E_0) = \left(1 - \frac{r_0}{a} \right) \sin \Delta E - \frac{\mathbf{r}_0 \cdot \dot{\mathbf{r}}_0}{\sqrt{\mu a}} \left(1 - \cos \Delta E \right).$$

Now substitute this result back into equation (5.47):

$$t - t_0 = \sqrt{\frac{a^3}{\mu}} \left[\Delta E - \left(1 - \frac{r_0}{a} \right) \sin \Delta E + \frac{\mathbf{r}_0 \cdot \dot{\mathbf{r}}_0}{\sqrt{\mu a}} \left(1 - \cos \Delta E \right) \right]. \tag{5.48}$$

This is Kepler's equation relating time and delta eccentric anomaly.

An equation for the distance can now be developed. From equation (4.12),

$$r = \sqrt{\frac{\mu}{a}} \frac{dt}{d(\Delta E)},$$

where we have used the equation

$$d(\Delta E) = d(E - E_0) = dE.$$

Therefore, from equation (5.48) for the time,

$$r = a \frac{d}{d(\Delta E)} \left[\Delta E - \left(1 - \frac{r_0}{a}\right) \sin \Delta E + \frac{\mathbf{r}_0 \cdot \dot{\mathbf{r}}_0}{\sqrt{\mu a}} (1 - \cos \Delta E) \right].$$

Performing the differentiation,

$$r = a \left[1 - \left(1 - \frac{r_0}{a}\right) \cos \Delta E + \frac{\mathbf{r}_0 \cdot \dot{\mathbf{r}}_0}{\sqrt{\mu a}} \sin \Delta E \right].$$

Note that only one orbital element (the semi-major axis) appears and the only limitation is that the orbit be elliptical ($a > 0$). Also, since only changes in eccentric anomaly are computed, the initial value of E is not needed.

We now summarize this formulation in algorithmic form.

5.3.1 Algorithm No. 3: Computation of Delta-E

Given $\mathbf{r}_0, \dot{\mathbf{r}}_0, t_0, \Delta E_1$ (initial guess for ΔE).
Find $\mathbf{r}$ and $\dot{\mathbf{r}}$ at time t.
Procedure

▷ Compute the magnitude of $\mathbf{r}_0$ and the semi-major axis,

$$r_0 = \|\mathbf{r}_0\|$$

$$a = \left(\frac{2}{r_0} - \frac{\dot{\mathbf{r}}_0 \cdot \dot{\mathbf{r}}_0}{\mu}\right)^{-1}.$$

　　If $a \leq 0$ STOP (not elliptical).
　　If $a > 0$ CONTINUE.
▷ Compute a new ΔE using the Newton-Raphson equations (compare with Algorithm No. 2),

$$F(\Delta E_1) = \Delta E_1 - \left(1 - \frac{r_0}{a}\right) \sin \Delta E_1 + \frac{\mathbf{r}_0 \cdot \dot{\mathbf{r}}_0}{\sqrt{\mu a}} (1 - \cos \Delta E_1)$$

$$- \sqrt{\frac{\mu}{a^3}} (t - t_0)$$

$$\left(\frac{dF}{d\Delta E}\right)_1 = 1 - \left(1 - \frac{r_0}{a}\right) \cos \Delta E_1 + \frac{\mathbf{r}_0 \cdot \dot{\mathbf{r}}_0}{\sqrt{\mu a}} \sin \Delta E_1$$

$$\Delta E = \Delta E_1 - \frac{F(\Delta E_1)}{(dF/d\Delta E)_1}.$$

▷ If $|\Delta E - \Delta E_1| >$ our error tolerance, then set $\Delta E_1 = \Delta E$ and go back
   to the second step.

▷ We have convergence for ΔE, so compute f, g, and $\mathbf{r}$ at time t,

$$f = 1 - \frac{a}{r_0}(1 - \cos \Delta E)$$

$$g = t - t_0 - \sqrt{\frac{a^3}{\mu}}(\Delta E - \sin \Delta E)$$

$$\mathbf{r} = f\, \mathbf{r}_0 + g\, \dot{\mathbf{r}}_0$$

$$r = \|\mathbf{r}\|.$$

▷ Compute $\dot{f}$, $\dot{g}$, and $\dot{\mathbf{r}}$ at time t,

$$\dot{f} = -\frac{\sqrt{\mu a}}{r r_0}\sin \Delta E$$

$$\dot{g} = 1 - \frac{a}{r}(1 - \cos \Delta E)$$

$$\dot{\mathbf{r}} = \dot{f}\, \mathbf{r}_0 + \dot{g}\, \dot{\mathbf{r}}_0.$$

End.

5.4 f and g Functions—Universal Variable

The purpose here is to transform the f and g functions in ΔE (or equivalently
in ΔH for hyperbolic orbits) to a new variable that is valid for all orbits. This
formulation follows that of Battin [4]. We introduce the functions $S(z)$ and
$C(z)$ defined by

$$S(z) = \frac{1}{3!} - \frac{z}{5!} + \frac{z^2}{7!} - \cdots, \qquad (5.49)$$

which is

$$S(z) = \frac{\sqrt{z} - \sin \sqrt{z}}{(\sqrt{z})^3}$$

when $z > 0$; and is

$$S(z) = \frac{\sinh \sqrt{-z} - \sqrt{-z}}{(\sqrt{-z})^3}$$

when $z < 0$. Also,

$$C(z) = \frac{1}{2!} - \frac{z}{4!} + \frac{z^2}{6!} - \ldots, \tag{5.50}$$

which is

$$C(z) = \frac{1 - \cos\sqrt{z}}{z}$$

when $z > 0$; and is

$$C(z) = \frac{\cosh\sqrt{-z} - 1}{-z}$$

when $z < 0$.

Note that from Appendix E (Stumpff functions) that

$$S(z) = c_3(z)$$

and

$$C(z) = c_2(z)$$

We can also easily verify the derivatives

$$\frac{dC(z)}{dz} = \frac{1}{2z}(1 - zS(z) - 2C(z))$$
$$\frac{dS(z)}{dz} = \frac{1}{2z}(C(z) - 3S(z))$$

from the derivative formulas given in Appendix E.

Now we will introduce these functions into the f, g, $\dot{f}$, and $\dot{g}$ functions as given by equations (5.42)–(5.44) and (5.46), as well as Kepler's equation (5.48). Recall that these equations involved the trigonometric functions $\sin \Delta E$ and $\cos \Delta E$. Note the expansion for $\cos \Delta E$,

$$\cos \Delta E = 1 - \frac{(\Delta E)^2}{2!} + \frac{(\Delta E)^4}{4!} - \frac{(\Delta E)^6}{6!} + - \ldots$$

Now define the auxiliary variable

$$\Delta E = \sqrt{\alpha_0}\, x \qquad \rightarrow \qquad x = \frac{\Delta E}{\sqrt{\alpha_0}},$$

where $\alpha_0 = 1/a$ and a is the semi-major axis. Note that $\alpha_0 > 0$ for elliptical orbits, $\alpha_0 < 0$ for hyperbolic orbits and $\alpha_0 = 0$ for parabolic orbits.

We will confine our analysis to the elliptical case for the moment. In terms of x, $\cos \Delta E$ becomes

$$\cos \Delta E = 1 - \frac{\alpha_0 x^2}{2!} + \frac{\alpha_0^2 x^4}{4!} - \frac{\alpha_0^3 x^6}{6!} + - \cdots,$$

and by factoring $\alpha_0 x^2$,

$$\cos \Delta E = 1 - \alpha_0 x^2 \left[\frac{1}{2!} - \frac{\alpha_0 x^2}{4!} + \frac{\alpha_0^2 x^4}{6!} - + \cdots \right].$$

But the term in the brackets is the same as $C(\alpha_0 x^2)$, as shown in equation (5.50) (where we let $z = \alpha_0 x^2$). So $\cos \Delta E$ becomes

$$\cos \Delta E = 1 - \alpha_0 x^2 C(\alpha_0 x^2). \tag{5.51}$$

We treat $\sin \Delta E$ in a similar way. The expansion for $\sin \Delta E$ is

$$\sin \Delta E = \Delta E - \frac{(\Delta E)^3}{3!} + \frac{(\Delta E)^5}{5!} - \frac{(\Delta E)^7}{7!} + - \cdots$$

By factoring ΔE once, we obtain

$$\sin \Delta E = \Delta E \left(1 - \frac{(\Delta E)^2}{3!} + \frac{(\Delta E)^4}{5!} - \frac{(\Delta E)^6}{7!} + - \cdots \right),$$

and once again,

$$\sin \Delta E = \Delta E \left(1 - (\Delta E)^2 \left[\frac{1}{3!} + \frac{(\Delta E)^2}{5!} - \frac{(\Delta E)^4}{7!} + - \cdots \right] \right).$$

Again using $\Delta E = \sqrt{\alpha_0} x$ the equation for $\sin \Delta E$ becomes

$$\sin \Delta E = \sqrt{\alpha_0} x \left(1 - \alpha_0 x^2 \left[\frac{1}{3!} + \frac{(\Delta E)^2}{5!} - \frac{(\Delta E)^4}{7!} + - \cdots \right] \right).$$

The term in the brackets is $S(\alpha_0 x^2)$ as shown in equation (5.49), where $z = \alpha_0 x^2$. We have therefore

$$\sin \Delta E = \sqrt{\alpha_0} x \left[1 - \alpha_0 x^2 S(\alpha_0 x^2) \right]. \tag{5.52}$$

Now recall equation (5.42),

$$f = 1 - \frac{a}{r_0} (1 - \cos \Delta E)$$

Substitution of (5.51) yields

$$f = 1 - \frac{x^2}{r_0} C(\alpha_0 x^2).$$ (5.53)

Similarly, from equations (5.43) and (5.52),

$$\dot{f} = -\frac{\sqrt{\mu a}}{r \, r_0} \sin \Delta E,$$

we obtain

$$\dot{f} = \frac{\sqrt{\mu}}{r \, r_0} \left[\alpha_0 x^3 S(\alpha_0 x^2) - x \right].$$ (5.54)

And from equations (5.44) and (5.51),

$$\dot{g} = 1 - \frac{a}{r} (1 - \cos \Delta E),$$

we obtain,

$$\dot{g} = 1 - \frac{x^2}{r} C(\alpha_0 x^2)$$ (5.55)

Also from equations (5.46) and (5.52),

$$g = t - t_0 - \sqrt{\frac{a^3}{\mu}} (\Delta E - \sin \Delta E),$$

we obtain

$$g = t - t_0 - \frac{1}{\sqrt{\mu}} x^3 S(\alpha_0 x^2)$$ (5.56)

In summary, the f and g functions and their derivatives formulated in the universal variable are

$$f = 1 - \frac{x^2}{r_0} C(\alpha_0 x^2)$$

$$\dot{f} = \frac{\sqrt{\mu}}{r \, r_0} \left[\alpha_0 x^3 S(\alpha_0 x^2) - x \right]$$

$$g = t - t_0 - \frac{1}{\sqrt{\mu}} x^3 S(\alpha_0 x^2)$$

$$\dot{g} = 1 - \frac{x^2}{r} C(\alpha_0 x^2).$$

Also recall equation (5.48) (Kepler's equation),

$$t - t_0 = \sqrt{\frac{a^3}{\mu}} \left[\Delta E - \left(1 - \frac{r_0}{a} \right) \sin \Delta E + \frac{\mathbf{r}_0 \cdot \dot{\mathbf{r}}_0}{\sqrt{\mu a}} \left(1 - \cos \Delta E \right) \right],$$

which, after using equations (5.51) and (5.52), becomes

$$t - t_0 = \frac{1}{\sqrt{\mu \alpha_0^3}} \left[\alpha_0^{\frac{3}{2}} x^3 S(\alpha_0 x^2) + r_0 \alpha_0 \sqrt{\alpha_0} x \left(1 - \alpha_0 x^2 S(\alpha_0 x^2) \right) \right.$$

$$\left. + \sqrt{\frac{\alpha_0}{\mu}} (\mathbf{r}_0 \cdot \dot{\mathbf{r}}_0) \alpha_0 x^2 C(\alpha_0 x^2) \right].$$

Note the cancelation of $\alpha_0^{\frac{3}{2}}$, which results in

$$\sqrt{\mu}(t - t_0) = \frac{\mathbf{r}_0 \cdot \dot{\mathbf{r}}_0}{\sqrt{\mu}} x^2 C(\alpha_0 x^2) + x^3 S(\alpha_0 x^2)(1 - r_0 \alpha_0) + r_0 x. \qquad (5.57)$$

Applying equations (5.51) and (5.52) to the equation for the distance in delta eccentric anomaly that was developed in §5.3 , we have

$$r = a \left[1 - \left(1 - \frac{r_0}{a} \right) \cos \Delta E + \frac{\mathbf{r}_0 \cdot \dot{\mathbf{r}}_0}{\sqrt{\mu a}} \sin \Delta E \right]$$

$$= \frac{1}{\alpha_0} \left[\alpha_0 x^2 C(\alpha_0 x^2) + r_0 \alpha_0 (1 - \alpha_0 x^2 C(\alpha_0 x^2)) \right.$$

$$\left. + \mathbf{r}_0 \cdot \dot{\mathbf{r}}_0 \sqrt{\frac{\alpha_0}{\mu}} \sqrt{\alpha_0} x(1 - \alpha_0 x^2 S(\alpha_0 x^2)) \right].$$

Canceling α_0 and rearranging we get

$$r = r_0 + x^2 C(\alpha_0 x^2)(1 - r_0 \alpha_0) + \frac{\mathbf{r}_0 \cdot \dot{\mathbf{r}}_0}{\sqrt{\mu}} x(1 - \alpha_0 x^2 S(\alpha_0 x^2)).$$

Alternatively, we can obtain the equation for the distance in the universal variable by expressing r in the delta eccentric anomaly as we did in §5.3:

$$r = \sqrt{\frac{\mu}{a}} \frac{dt}{d(\Delta E)},$$

but by definition

$$\Delta E = \sqrt{\alpha_0} x.$$

Substituting into the equation for r,

$$r = \sqrt{\mu \alpha_0} \, \frac{dt}{d(\sqrt{\alpha_0} x)} = \sqrt{\mu} \, \frac{dt}{dx},$$

where dt/dx is obtained by differentiation of (5.57). This approach will be useful in the iteration procedure for x.

The equations (5.53)–(5.57) represent a solution of the form

$$\mathbf{r} = f \, \mathbf{r}_0 + g \, \dot{\mathbf{r}}_0$$

$$\dot{\mathbf{r}} = \dot{f} \, \mathbf{r}_0 + \dot{g} \, \dot{\mathbf{r}}_0$$

to the two-body problem when the initial conditions are given. Although they were developed for elliptical orbits ($\alpha_0 > 0$), they can be shown to be valid for the parabolic ($\alpha_0 = 0$) and hyperbolic ($\alpha_0 < 0$) orbits. In fact, the two-body differential equation of motion,

$$\ddot{\mathbf{r}} + \frac{\mu}{r^3}\mathbf{r} = 0,$$

is satisfied by direct substitution of $\mathbf{r}$ and $\dot{\mathbf{r}}$ given in the terms of x through equations (5.53)–(5.57), with no assumptions made about the sign of α_0.

5.4.1 Algorithm No. 4: Using Universal Variables

Given $\mathbf{r}_0$, $\dot{\mathbf{r}}_0$ and t_0.
Find $\mathbf{r}$ and $\dot{\mathbf{r}}$ at time t.
Procedure

▷ First compute the magnitude of $\mathbf{r}_0$ and α_0:

$$r_0 = \|\mathbf{r}_0\|$$

$$\alpha_0 = \frac{2}{r_0} - \frac{\dot{\mathbf{r}}_0 \cdot \dot{\mathbf{r}}_0}{\mu}.$$

▷ Next iterate Kepler's equation for x (see Algorithm No. 5 in the next section).

▷ Finally compute $\mathbf{r}$ and $\dot{\mathbf{r}}$:

$$f = 1 - \frac{x^2}{r_0} C(\alpha_0 x^2)$$

$$g = t - t_0 - \frac{x^3}{\sqrt{\mu}} S(\alpha_0 x^2)$$

$$\mathbf{r} = f\,\mathbf{r}_0 + g\,\dot{\mathbf{r}}_0$$

$$r = \|\mathbf{r}\|$$

$$\dot{f} = \frac{\sqrt{\mu}}{r r_0}\left[\alpha_0 x^3\,S(\alpha_0 x^2) - x\right]$$

$$\dot{g} = 1 - \frac{x_2}{r}\,C(\alpha_0 x^2)$$

$$\dot{\mathbf{r}} = \dot{f}\,\mathbf{r}_0 + \dot{g}\,\dot{\mathbf{r}}_0.$$

End.

5.4.2 Algorithm No. 5: Kepler's Equation—Solution

Given t_0, t, r_0, α_0 and x_1 (first guess for x).
Find $x, C(\alpha_0 x^2)$ and $S(\alpha_0 x^2)$.
Procedure

▷ Calculate the current value of x,

$$z_1 = \alpha_0 x_1^2$$

$$S(z_1) = \frac{1}{3!} - \frac{z_1}{5!} + \frac{z_1^2}{7!} - \cdots$$

$$C(z_1) = \frac{1}{2!} - \frac{z_1}{4!} + \frac{z_1^2}{6!} - \cdots$$

$$F(x_1) = \sqrt{\mu}(t - t_0) - \left[\frac{\mathbf{r}_0 \cdot \dot{\mathbf{r}}_0}{\sqrt{\mu}}x_1^2 C(z_1) + x_1^3 S(z_1)(1 - r_0\alpha_0) + r_0 x_1\right]$$

$$\left(\frac{dF}{dx}\right)_1 = -\frac{\mathbf{r}_0 \cdot \dot{\mathbf{r}}_0}{\sqrt{\mu}}(x_1 - z_1 x_1 S(z_1)) - (1 - r_0\alpha_0)x_1^2 C(z_1) - r_0$$

$$x = x_1 - \frac{F(x_1)}{(dF/dx)_1}.$$

▷ If $|x - x_1| >$ our error tolerance, then set $x_1 = x$ and go back to the previous step.

▷ If x is within our error tolerance, then the value of x has been found. Calculate the final values for $C(\alpha_0 x^2)$ and $S(\alpha_0 x^2)$.

End.

5.5 f and g Functions in Time

The f and g functions can be considered as explicit functions of time (Bond [13]). If we substitute equations (5.28) and (5.29) into the differential equation of motion of the two-body problem (eq. 2.8) and equate coefficients of the initial conditions ($\mathbf{r}_0$ and $\dot{\mathbf{r}}_0$), we obtain the two scalar differential equations

$$\ddot{f} + q\,f = 0 \tag{5.58}$$

$$\ddot{g} + q\,g = 0. \tag{5.59}$$

We have introduced the function

$$q = \frac{\mu}{r^3} \tag{5.60}$$

from which we obtain the differential equation,

$$r\,\dot{q} + 3q\,\dot{r} = 0. \tag{5.61}$$

The differential equation for the distance r is

$$\ddot{r} = \frac{c^2}{r^3} - \frac{\mu}{r^2}, \tag{5.62}$$

which we obtained by twice differentiating equation (3.3) and using equation (3.7). Using the definition for q, the differential equation (5.62) becomes

$$\ddot{r} = q\left(\frac{c^2}{\mu} - r\right). \tag{5.63}$$

The f and g functions as well as the function q and the distance r can be expanded in a Taylor's series in time,

$$f = \sum_{n=0}^{\infty} a_n(t - t_0)^n \tag{5.64}$$

$$g = \sum_{n=0}^{\infty} b_n(t - t_0)^n \tag{5.65}$$

$$q = \sum_{n=0}^{\infty} c_n(t - t_0)^n \tag{5.66}$$

$$r = \sum_{n=0}^{\infty} d_n(t - t_0)^n. \tag{5.67}$$

The procedure now is to substitute the four series given by equations (5.64)–(5.67) into the four differential equations (5.58), (5.59), (5.61), and (5.62), and then solve for the coefficients a_n, b_n, c_n, and d_n by comparison of the coefficients of powers of time. The central mathematical device used is the general relation

$$\left(\sum_{n=0}^{\infty} \alpha_n x^n\right)\left(\sum_{n=0}^{\infty} \beta_n x^n\right) = \sum_{n=0}^{\infty} \sum_{v=0}^{n} \alpha_v \beta_{n-v} x^n, \tag{5.68}$$

which converts the product of two infinite series to a double summation. The resulting recurrence relations are

$$d_{n+2} = \frac{1}{(n+1)(n+2)}\left(\frac{c^2}{\mu} c_n - \sum_{v=0}^{n} c_{n-v} d_v\right) \tag{5.69}$$

$$c_n = \frac{-1}{n\,d_0}\left[3 c_0 n\, d_n + \sum_{v=1}^{n-1} v(3\, c_{n-v} d_v + c_v d_{n-v})\right] \tag{5.70}$$

$$a_{n+2} = \frac{-1}{(n+1)(n+2)} \sum_{v=0}^{n} c_v a_{n-v} \tag{5.71}$$

$$b_{n+2} = \frac{-1}{(n+1)(n+2)} \sum_{v=0}^{n} c_v b_{n-v}. \tag{5.72}$$

The starting values for the coefficients in the f and g series are

$$f(t_0) = a_0 = 1$$
$$\dot{f}(t_0) = a_1 = 0$$
$$g(t_0) = b_0 = 0$$
$$\dot{g}(t_0) = b_1 = 1,$$

which can be verified directly from equations (5.28) and (5.29).

The starting values for the coefficients in the series for q and r from equations (5.60) and (5.67) are

$$q(t_0) = c_0 = \frac{\mu}{r_0^3}$$
$$r(t_0) = d_0 = r_0$$
$$\dot{r}(t_0) = d_1 = \dot{r}_0.$$

The advantage of the solution for $\mathbf{r}$ and $\dot{\mathbf{r}}$ is that there is no need to solve Kepler's equation. The disadvantage is that the series fails to converge beyond a certain value of time (ρ), the radius of convergence. Fortunately, the radius of convergence can be calculated from Moulton's formulas, which are given in [12]:

(1) For an ellipse

$$\rho = \sqrt{\frac{a^3}{\mu}} \left(M_0^2 + \left[\ln \left(\frac{1 + \sqrt{1 - e^2}}{e} \right) - \sqrt{1 - e^2} \right]^2 \right)^{1/2},$$

where

$$M_0 = E_0 - e \sin E_0 = \sqrt{\frac{\mu}{a^3}} \, (t_0 - t_\pi).$$

(2) For a parabola

$$\rho = \sqrt{\frac{p^3}{3\mu}} \left[1 + \frac{9\mu}{p^3} (t_0 - t_\pi)^2 \right]^{1/2}.$$

(3) For a hyperbola

$$\rho = \sqrt{\frac{-a^3}{\mu}} \left[N_0^2 + (\tan \alpha - \alpha)^2 \right]^{1/2},$$

where

$$N_0 = \sqrt{\frac{-\mu}{a^3}} \, (t_0 - t_\pi)$$

$$\alpha = \cos^{-1} \left(\frac{1}{e} \right).$$

6

TWO-POINT BOUNDARY VALUE PROBLEMS

6.1 Introduction

Previously we have discussed the solution of the two-body problem where we are given the initial conditions $\mathbf{r}_0$ and $\dot{\mathbf{r}}_0$ of the mass m_2 with respect to the mass m_1. With the initial conditions specified we have given several methods for finding the solution at any other time. When conditions are given, or measured, at various instants of time along the trajectory, we refer to the numerical process of finding a solution as *orbit determination*. An early successful instance of orbit determination was in 1801 by Gauss, who used the observations of the astronomer Piazza to compute the orbit of the minor planet Ceres. The data that Gauss used were of angles only; that is, three unit vectors at three times.

In the two methods discussed below we will assume that we are given the initial and final position vectors. The first method is called "Lambert's problem," in which we specify the time of flight between the two position vectors. The second method is unconstrained in time, though a constraint is imposed on the components of the final velocity.

6.2 Lambert's Problem

Lambert's problem is defined as follows: Give two position vectors $\mathbf{r}_0$ and $\mathbf{r}$ at times t_0 and t of the mass m_2 with respect to m_1; find the solution for the trajectory connecting the two positions. It is sufficient to solve for the initial velocity $\dot{\mathbf{r}}_0$ only in order to obtain the solution, since we can use $\mathbf{r}_0$ and $\dot{\mathbf{r}}_0$ at t_0 to calculate $\mathbf{r}$ and $\dot{\mathbf{r}}$ at any other time t (refer to §4.5 and Algorithm No. 1 of chapter 4). Lambert's problem is basic to mission planning, targeting, guidance, and rendezvous.

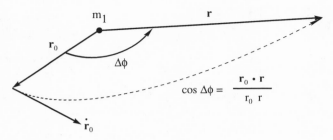

Figure 6.1   Lambert's problem.

We now consider the most *basic* approach to the solution of this problem as given by Bate, Mueller, and White [3], confining ourselves to elliptical orbits. For the *initial value* problem (given r_0, $\dot{r}_0$, t_0, and t), we have the f and g solution

$$\mathbf{r} = f\,\mathbf{r}_0 + g\,\dot{\mathbf{r}}_0$$

$$\dot{\mathbf{r}} = \dot{f}\,\mathbf{r}_0 + \dot{g}\,\dot{\mathbf{r}}_0,$$

where f, g, $\dot{f}$, and $\dot{g}$ are given by equations (5.37)–(5.40) for the true anomaly form and by equations (5.42)–(5.44) and (5.46) for the eccentric anomaly form. These forms are numerically the same.

$$f = 1 - \frac{r}{p}(1 - \cos \Delta\phi) = 1 - \frac{a}{r_0}(1 - \cos \Delta E) \tag{6.1}$$

$$g = \frac{rr_0}{\sqrt{\mu p}} \sin \Delta\phi = t - t_0 - \sqrt{\frac{a^3}{\mu}}(\Delta E - \sin \Delta E) \tag{6.2}$$

$$\dot{f} = \sqrt{\frac{\mu}{p}} \left(\frac{1 - \cos \Delta\phi}{\sin \Delta\phi} \right) \left[\frac{1}{p}(1 - \cos \Delta\phi) - \frac{1}{r_0} - \frac{1}{r} \right]$$

$$= -\frac{\sqrt{\mu a}}{rr_0} \sin \Delta E \tag{6.3}$$

$$\dot{g} = 1 - \frac{r_0}{p}(1 - \cos \Delta\phi) = 1 - \frac{a}{r}(1 - \cos \Delta E). \tag{6.4}$$

Note that equations (6.1) and (6.4) give the same result:

$$rr_0(1 - \cos \Delta\phi) = ap(1 - \cos \Delta E).$$

Also note that r_0, r, and $\Delta\phi$ are known. So we have *three equations*—(6.1), (6.2), and (6.3)—which can be solved for the *three unknowns* a, p, and ΔE.

We set up the following procedure

Given   The initial and final positions r_0 and r, and the time of flight Δt ($= t - t_0$).
Find   The initial velocity vector $\dot{r}_0$.
Procedure

  ▷ Choose an initial guess for ΔE. One guess could be setting $\Delta E = \Delta\phi$.
  ▷ Adjust ΔE by numerical iteration.
  ▷ Compute a and p by solving equations (6.1) and (6.3),

$$ap(1 - \cos \Delta E) = rr_0(1 - \cos \Delta\phi)$$

$$\sqrt{\frac{\mu}{p}} \left(\frac{1 - \cos \Delta\phi}{\sin \Delta\phi} \right) \left[\frac{1}{p}(1 - \cos \Delta\phi) - \frac{1}{r_0} - \frac{1}{r} \right] = -\frac{\sqrt{\mu a}}{rr_0} \sin \Delta E,$$

simultaneously (two equations, two unknowns) using the current value of ΔE.

  ▷ Compute $\tilde{\Delta} t$ from equation (6.2),

$$\tilde{\Delta} t = \sqrt{\frac{a^3}{\mu}}(\Delta E - \sin \Delta E) + \frac{rr_0}{\sqrt{\mu p}} \sin \Delta\phi.$$

  ▷ If $\left| \tilde{\Delta} t - \Delta t \right| >$ our error tolerance, then go back to the second step.
  ▷ Compute f, g and $\dot{r}_0$,

$$f = 1 - \frac{r}{p}(1 - \cos \Delta\phi)$$

$$g = \frac{rr_0}{\sqrt{\mu p}} \sin \Delta\phi$$

$$\dot{r}_0 = \frac{1}{g}(r - f\, r_0).$$

End.

6.3 Universal Variable

The formulation of Lambert's problem in the universal variable we present is taken from Battin [4], who in turn gives credit to John Deyst of the MIT Instrumentation Laboratory. Battin in reference [4] uses the designation "Lambert's problem," while Bate, Mueller, and White [3] refer to it as the "Gauss problem." We will follow Battin's designation.

In this approach we equate the functions f, g, $\dot{f}$, and $\dot{g}$ in the true anomaly (eqs. 5.37–5.40) to the corresponding functions in the universal variable (eqs. 5.53–5.56),

$$f = 1 - \frac{r}{p}(1 - \cos \Delta\phi) = 1 - \frac{x^2}{r_0}C(\alpha_0 x^2). \tag{6.5}$$

$$g = \frac{rr_0}{\sqrt{\mu p}} \sin \Delta\phi = t - t_0 - \frac{x^3}{\sqrt{\mu}}S(\alpha_0 x^2) \tag{6.6}$$

$$\dot{f} = \sqrt{\frac{\mu}{p}} \left(\frac{1 - \cos \Delta\phi}{\sin \Delta\phi} \right) \left[\frac{1}{p}(1 - \cos \Delta\phi) - \frac{1}{r_0} - \frac{1}{r} \right]$$

$$= \frac{\sqrt{\mu}}{rr_0} \left[\alpha_0 x^3 S(\alpha_0 x^2) - x \right] \tag{6.7}$$

$$\dot{g} = 1 - \frac{r_0}{p}(1 - \cos \Delta\phi) = 1 - \frac{x^2}{r}C(\alpha_0 x^2). \tag{6.8}$$

Note that equations (6.5) and (6.8) reduce to the same result:

$$\frac{r}{p}(1 - \cos \Delta\phi) = \frac{x^2}{r_0}C(\alpha_0 x^2). \tag{6.9}$$

We have three equations for three unknowns α_0, p, and x. Solve equation (6.9) for x, ignoring the dependence of $C(\alpha_0 x^2)$ on x.

$$x = \sqrt{\frac{rr_0}{pC}(1 - \cos \Delta\phi)}, \tag{6.10}$$

where $C = C(\alpha_0 x^2)$. Also let $S = S(\alpha_0 x^2)$ below. Now substitute equation (6.10) into the right side of equation (6.7):

$$\sqrt{\frac{\mu}{p}} \left(\frac{1 - \cos \Delta\phi}{\sin \Delta\phi} \right) \left[\frac{1}{p}(1 - \cos \Delta\phi) - \frac{1}{r_0} - \frac{1}{r} \right]$$

$$= \frac{\sqrt{\mu}}{rr_0} \sqrt{\frac{rr_0}{pC}}(1 - \cos \Delta\phi) \left[\alpha_0 x^2 S - 1 \right].$$

Note $\sqrt{\mu/p}$ cancels on each side. Multiply by $\frac{rr_0 \sin \Delta\phi}{1 - \cos \Delta\phi}$ and rearrange slightly to get

$$\frac{rr_0}{p}(1 - \cos \Delta\phi) = r_0 + r - \sin \Delta\phi \sqrt{\frac{rr_0}{1 - \cos \Delta\phi}} \left(\frac{1 - \alpha_0 x^2 S}{\sqrt{C}} \right). \tag{6.11}$$

Now on the left side of (6.11) let

$$y = \frac{rr_0}{p}(1 - \cos \Delta\phi) \tag{6.12}$$

and on the right side of (6.11) let

$$A = \sin \Delta\phi \sqrt{\frac{rr_0}{1 - \cos \Delta\phi}}, \tag{6.13}$$

noting that A is constant. Then equation (6.11) becomes

$$y = r_0 + r - A\left(\frac{1 - zS}{\sqrt{C}}\right), \tag{6.14}$$

where, as in Appendix E, we have used

$$z = \alpha_0 x^2. \tag{6.15}$$

Now use equation (6.12) in (6.10):

$$x = \frac{1}{\sqrt{C}}\sqrt{\frac{rr_0}{p}(1 - \cos \Delta\phi)} \tag{6.16}$$

$$x = \sqrt{\frac{y}{C}}. \tag{6.17}$$

Now from equation (6.6) solve for $t - t_0$,

$$\sqrt{\mu}(t - t_0) = \frac{rr_0 \sin \Delta\phi}{\sqrt{p}} + x^3 S, \tag{6.18}$$

but from equation (6.13),

$$\sin \Delta\phi = A\sqrt{\frac{1 - \cos \Delta\phi}{rr_0}},$$

and from equation (6.12),

$$\frac{1}{\sqrt{p}} = \sqrt{\frac{y}{rr_0(1 - \cos \Delta\phi)}},$$

so equation (6.18) reduces to

$$\sqrt{\mu}(t - t_0) = A\sqrt{y} + x^3 S. \tag{6.19}$$

Now from equations (6.12) and (6.13) the equations for f, g, and $\dot{g}$ (eqs. 6.5, 6.6, and 6.8) assume the simple form

$$f = 1 - \frac{y}{r_0} \tag{6.20}$$

$$g = A\sqrt{\frac{y}{\mu}} \tag{6.21}$$

$$\dot{g} = 1 - \frac{y}{r}. \tag{6.22}$$

The *initial velocity* ($\dot{\mathbf{r}}_0$) can be found since

$$\mathbf{r} = \mathbf{r}_0 f + \dot{\mathbf{r}}_0 g.$$

Solve for $\dot{\mathbf{r}}_0$:

$$\dot{\mathbf{r}}_0 = \frac{1}{g}(\mathbf{r} - \mathbf{r}_0 f). \tag{6.23}$$

The *final velocity* ($\dot{\mathbf{r}}$) can also be found since

$$\dot{\mathbf{r}} = \mathbf{r}_0 \dot{f} + \dot{\mathbf{r}}_0 \dot{g}.$$

Substitute (6.23):

$$\dot{\mathbf{r}} = \mathbf{r}_0 \dot{f} + \dot{g}\frac{1}{g}(\mathbf{r} - \mathbf{r}_0 f).$$

Factor $1/g$ on the right side:

$$\dot{\mathbf{r}} = \frac{1}{g}\left[\mathbf{r}_0(g\dot{f} - \dot{g}f) + \dot{g}\,\mathbf{r}\right].$$

Now use $f\dot{g} - g\dot{f} = 1$ to get

$$\dot{\mathbf{r}} = \frac{1}{g}(\dot{g}\,\mathbf{r} - \mathbf{r}_0). \tag{6.24}$$

We now set up the solution algorithm.

6.3.1 Algorithm No. 6: Solution of Lambert's Problem

Given $\mathbf{r}_0$, $\mathbf{r}$, and Δt $(= t - t_0)$.
Find $\dot{\mathbf{r}}_0$ and $\dot{\mathbf{r}}$.

Procedure

▷ Compute r, r_0, $\Delta\phi$:

$$r = \|\mathbf{r}\|$$

$$r_0 = \|\mathbf{r}_0\|$$

$$\Delta\phi = \begin{cases} \cos^{-1} \frac{\mathbf{r}_0 \cdot \mathbf{r}}{r_0 r} \\ \text{or} \\ 2\pi - \cos^{-1} \frac{\mathbf{r}_0 \cdot \mathbf{r}}{r_0 r} \end{cases}$$

(*Note:* See Appendix D for details on the proper choice of $\Delta\phi$.) From (6.13),

$$A = \sqrt{\frac{r r_0}{1 - \cos \Delta\phi}} \sin \Delta\phi.$$

▷ Guess or adjust the value of z (for example, an iteration using eq. 6.19).
▷ Compute $C(z)$ and $S(z)$ from the series given by equations (5.49) and (5.50):

$$C = \frac{1}{2!} - \frac{z}{4!} + \frac{z^2}{6!} - \frac{z^3}{8!} + - \cdots$$

$$S = \frac{1}{3!} - \frac{z}{5!} + \frac{z^2}{7!} - \frac{z^3}{9!} + - \cdots.$$

▷ Compute y and x from equations (6.14) and (6.17):

$$y = r_0 + r - A\left[\frac{1 - zS}{\sqrt{C}}\right]$$

$$x = \sqrt{\frac{y}{C}}.$$

▷ Compute the trial value of Δt from equation (6.19):

$$\tilde{\Delta t} = \frac{1}{\sqrt{\mu}}\left[x^3 S + A\sqrt{y}\right]. \tag{6.25}$$

▷ If $\left|\Delta t - \tilde{\Delta t}\right| >$ our error tolerance, then go back to the second step.
▷ If the value for $\tilde{\Delta t}$ is within our error tolerance, then the process has converged, so compute $\dot{\mathbf{r}}_0$ and $\dot{\mathbf{r}}$ using equations (6.20) through (6.24):

$$f = 1 - \frac{y}{r_0}$$

$$g = A\sqrt{\frac{y}{\mu}}$$

$$\dot{g} = 1 - \frac{y}{r}$$

$$\dot{\mathbf{r}}_0 = \frac{1}{g}[\mathbf{r} - f\,\mathbf{r}_0]$$

$$\dot{\mathbf{r}} = \frac{1}{g}[\dot{g}\,\mathbf{r} - \mathbf{r}_0]$$

End.

6.4 Linear Terminal Velocity Constraint

This problem is usually referred to by the acronym LTVCON and is defined as follows:

Given the initial and terminal position vectors $\mathbf{r}_0$ and $\mathbf{r}$, and the constraint on the terminal radial velocity,

$$\dot{r} = c_1 + c_2 v_{H},$$

find the initial velocity vector $\mathbf{v}_0$.

This is a generalization of the development of equation (4.14) in Battin [4]. Only internal references to this important targeting algorithm have been found. Dave Long [51] states that in the early 1970s, E. C. Lineberry of NASA-JSC told him that Battin's equation (4.14) could be generalized into a targeting algorithm if a linear constraint between the radial and horizontal velocity components were imposed. Mr. Long credits W. Templeman with the earliest solution, in March 1974, in Charles Stark Draper Laboratory (CSDL) Report 10E-74-13. This was followed by NASA-JSC memorandum FM73(74-128) dated 26 June 1974 by Dave Long and Leroy McHenry on the idea and later by another CSDL Report (10E-74-53) written by Stanley Shepperd on 16 September 1974.

We begin with the position and velocity equations in terms of the f, g, $\dot{f}$, and $\dot{g}$ functions,

$$\mathbf{r} = \mathbf{r}_0\,f + \mathbf{v}_0\,g \tag{6.26}$$

$$\mathbf{v} = \mathbf{r}_0\,\dot{f} + \mathbf{v}_0\,\dot{g}, \tag{6.27}$$

which are equations (5.28) and (5.29) with $\mathbf{v}_0 = \dot{\mathbf{r}}_0$. The f, g, $\dot{f}$, and $\dot{g}$ functions are given by equations (5.37)–(5.40). Solve equation (6.26) for $\mathbf{v}_0$ to get

$$\mathbf{v}_0 = \frac{1}{g}\,(\mathbf{r} - \mathbf{r}_0\, f)\,,$$

which becomes, using equations (5.33) and (5.34),

$$\mathbf{v}_0 = \frac{\sqrt{\mu p}}{r\, r_0}\,\frac{1}{\sin \Delta\phi}\left[\mathbf{r} - \left(1 - \frac{r}{p}(1 - \cos \Delta\phi)\right)\mathbf{r}_0\right]. \tag{6.28}$$

Note that the only unknown in this equation is p. At the terminal point the horizontal velocity is found from equations (3.6) and (3.7) (in the discussion of Kepler's second law). The horizontal velocity is

$$v_{_H} = r\dot{\phi} = \frac{1}{r}\left(r^2\dot{\phi}\right) = \frac{1}{r}c = \frac{1}{r}\sqrt{\mu p}.$$

Thus the semi-latus rectum is

$$p = \frac{r^2 v_{_H}^2}{\mu}. \tag{6.29}$$

Now we must obtain an expression for $v_{_H}$. Since the terminal constraint is given by the terminal radial velocity ($\dot{r}$) component, we first derive an expression for $\dot{r}$. From (6.24),

$$\mathbf{v} = \frac{1}{g}\,(\dot{g}\,\mathbf{r} - \mathbf{r}_0)\,. \tag{6.30}$$

The terminal radial velocity is

$$\dot{r} = \hat{\mathbf{r}} \cdot \mathbf{v} = \frac{1}{g}\left(\dot{g}r - r_0\,\hat{\mathbf{r}} \cdot \hat{\mathbf{r}}_0\right)$$

or

$$\dot{r} = \frac{1}{g}\,(\dot{g}r - r_0 \cos \Delta\phi)\,. \tag{6.31}$$

Now we insert the constraint, $\dot{r} = c_1 + c_2 v_{_H}$, on the left side; insert the expressions for g and $\dot{g}$ from equations (5.34) and (5.35); and use equation (6.29) for p wherever it occurs. After some algebra we obtain the quadratic

$$A v_{_H}^2 - 2B v_{_H} - C = 0, \tag{6.32}$$

where

$$A = 2(1 - c_2 W) + K(W^2 + 1)$$

$$B = c_1 W$$

$$C = \frac{2\mu}{r}$$

$$K = \frac{r - r_0}{r_0}$$

$$W = \frac{\sin \Delta\phi}{1 - \cos \Delta\phi}.$$

Note that the quadrant for $\Delta\phi$ can be computed as given in Appendix D.
Equation (6.32) has the two solutions

$$v_H = \frac{1}{A}\left[B \pm \sqrt{B^2 + AC} \right],$$

where

$$B^2 + AC > 0$$

in order for a real solution to exist. Since $v_H > 0$, we choose

$$v_H = \frac{1}{A}\left[B + \sqrt{B^2 + AC} \right]$$

or, equivalently,

$$v_H = \frac{C}{-B + \sqrt{B^2 + AC}}. \tag{6.33}$$

Note that if the negative root were chosen, then the magnitude of the horizontal velocity (v_H) would be negative for the special case $c_1 = 0$. The value of p is found from equation (6.29) using the solution of equation (6.32). Since p is the only unknown in equation (6.28), we have a solution for $\mathbf{v}_0$.

7

APPLICATIONS

7.1 Interplanetary Trajectories

The precise computation of a trajectory in the solar system requires that the initial conditions are known and the perturbations that are mainly due to the planets are properly accounted for. A special perturbation method such as the one we will describe in a later section must be used for this computation. A second requirement for trajectories between planets, such as those that are flown by artificial spacecraft, is that the boundary conditions are given at both the initial and final times.

These two requirements present a great obstacle to practical interplanetary mission design. However, much of this burden is relieved by taking advantage of the fact that when the spacecraft is a large distance from a planet the trajectory can be considered as being a two-body problem involving the Sun and the spacecraft. In addition, when the spacecraft is close to a planet, the trajectory can also be considered as a two-body problem involving the planet and the spacecraft.

In the following sections we will describe an approach (Bond and Henry [20]) to the approximate solution of a trajectory that departs the vicinity of one planet and arrives in the vicinity of another planet. Notation is important for the understanding of this approach. We summarize the notation as follows:

1. Capital letters will apply to the trajectory when it is with respect to the Sun. For example,

   R is the position of the spacecraft with respect to the Sun;
   V is the velocity of the spacecraft with respect to the Sun;
   U is the velocity of the planet with respect to the Sun.

2. Lower-case letters will apply to the trajectory when it is with respect to a planet. For example, **r** and **v** are the position and velocity of the spacecraft with respect to a planet.

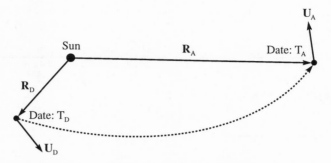

Figure 7.1   Trajectory of the spacecraft with respect to the sun.

3. The subscript D will refer to conditions at departure and the subscript A will refer to conditions at arrival.

7.1.1 Heliocentric Phase

Given the date of departure (T_D) from planet D and the date of arrival (T_A) at planet A, and if we assume the two planets to be massless, we can compute a conic trajectory between planet D and planet A. Since the ephemerides of the planets are given [75] as a function of date, we can obtain the positions ($\mathbf{R}_D$ and $\mathbf{R}_A$) and the velocities ($\mathbf{U}_D$ and $\mathbf{U}_A$) of the two planets at the dates T_D and T_A. Refer to figure 7.1.

Next we compute the conic between these two position vectors ($\mathbf{R}_D$ and $\mathbf{R}_A$) as we did in §6.3 when we presented Lambert's problem in universal variables (Algorithm No. 6). In this application of the algorithm, the initial and final position vectors are the position vectors $\mathbf{R}_D$ and $\mathbf{R}_A$, and the flight time (Δt) is the difference between the dates T_D and T_A. The solution of Lambert's problem yields the initial and final velocity vectors, $\mathbf{V}_D$ and $\mathbf{V}_A$, in a heliocentric coordinate system, which the coordinate system used for the planetary ephemerides. The velocity vectors with respect to the departure and arrival planets are

$$v_D = \mathbf{V}_D - \mathbf{U}_D$$
$$v_A = \mathbf{V}_A - \mathbf{U}_A,$$

and these are shown in figure 7.2.

The values v_D and v_A are referred to as the *v-infinity* (v_∞), or hyperbolic excess velocity vectors at departure and arrival. We can consider the v_∞ vectors as the interfaces between the heliocentric and planetocentric phases.

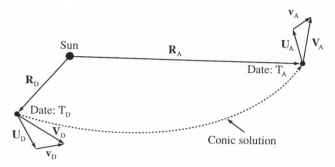

Figure 7.2 Velocity vectors for the planetary transfer problem.

The computation of the $\mathbf{v}_\infty$ vectors is presented in the following algorithm.

7.1.2 Algorithm No. 7: V-Infinity Vectors Solution

Given   The date of departure (T_D) from planet D and the date of arrival (T_A) at planet A.

Find   The v-infinity vectors $\mathbf{v}_D$ and $\mathbf{v}_A$ for each planet.

Procedure

▷ Using T_D, compute $\mathbf{R}_D$ and $\mathbf{U}_D$ from the planetary ephemeris.

▷ Using T_A, compute $\mathbf{R}_A$ and $\mathbf{U}_A$ from the planetary ephemeris.

▷ Using $T_A - T_D$ for Δt and $\mathbf{R}_D$ and $\mathbf{R}_A$ for the initial and final position vectors, use Algorithm No. 6 to compute the initial $(\mathbf{V}_D)$ and final $(\mathbf{V}_A)$ velocities of the spacecraft with respect to the Sun.

▷ Compute the $\mathbf{v}_\infty$ vectors:

$$\mathbf{v}_D = \mathbf{V}_D - \mathbf{U}_D$$

$$\mathbf{v}_A = \mathbf{V}_A - \mathbf{U}_A.$$

End.

7.1.3 Planetocentric Phase

The $\mathbf{v}_\infty$ vector is the asymptotic velocity of the spacecraft with respect to the planet as it departs from (or arrives at) the planet. The trajectory is always hyperbolic $(h > 0)$. The planet is assumed to have mass but the *Sun* is considered *massless*. We can make the following computations with respect to each planet, at departure and arrival.

From the conservation of energy,

$$h = -\frac{\mu}{2a} = \frac{1}{2}v^2 - \frac{\mu}{r},$$

where μ is the gravitational parameter of the planet. As $r \to \infty$, $v \to v_\infty$. For the planetary transfer problem $\mathbf{v}_\infty = \mathbf{v}_D$ or $\mathbf{v}_A$. We have, therefore,

$$h = -\frac{\mu}{2a} = \frac{1}{2}v_\infty^2$$

or

$$a = -\frac{\mu}{v_\infty^2}.$$

The $\mathbf{v}_\infty$ vector provides three constants of the motion: a direction in space $(\hat{\mathbf{S}})$ and the energy of the planetocentric trajectory (see fig. 7.3).

If we also *specify* r_π (the pericenter distance or radius of closest approach), we can compute the eccentricity. At the pericenter the conic equation gives

$$r_\pi = a(1 - e).$$

Therefore we can compute

$$e = 1 - \frac{r_\pi}{a}.$$

Since $a < 0$ for a hyperbola, we see that $e > 1$. Also knowing r_π, thus e, we can compute the limiting value of the true anomaly.

From the equation of a conic,

$$1 + e \cos \phi = \frac{a(1 - e^2)}{r}.$$

As $r \to \infty$, the true anomaly ϕ approaches its limiting value ϕ_L, which is seen from the above equation to be

$$\cos \phi_L = -\frac{1}{e}.$$

Since the eccentricity is positive, $90° < \phi_L < 270°$.

7.1.4 Planetary Flyby (Gravity Turn)

When a satellite (spacecraft, asteroid, comet, etc.) approaches a planet, the satellite is perturbed by the mass of the planet. The computation of the satellite's trajectory can be approximated by considering the trajectory of two conics, one conic *before* closest approach to the planet and the second *after* closest

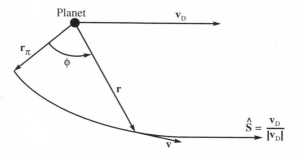

Figure 7.3   Trajectory of the spacecraft at departure.

approach. The perturbation due to the planet's mass causes an instantaneous velocity change. For precision work, this problem must be solved by a special perturbation method as discussed in chapter 9. The expression "gravity turn" is given to this problem since the gravity of the perturbing planet "turns" the path of the satellite in space. The expression is somewhat misleading since the planet's gravity generally changes all of the orbital elements of the satellite. During the Apollo program, where the Moon was the perturbing planet, another expression for this effect was "free return." The early Apollo missions were designed such that if the rocket engines (which were to place the spacecraft in an orbit about the Moon) failed, the spacecraft's orbit would be "bent" (perturbed) by the Moon into an orbit that returned the spacecraft to Earth. The expression "free flyby" is also frequently used. In this chapter we will make no distinction between these various expressions.

The gravity turn problem is defined as follows: The spacecraft (refer to fig. 7.4) departs planet D on date T_D, arrives at target planet T on date T_T, and continues to the final planet F, arriving on date T_F. Three restrictions on the problem are as follows:

1. No artificial (that is, use of rocket engine) impulsive velocity is allowed after departure.
2. Collision with the target planet is (obviously) not allowed.
3. The v-infinity vector magnitude ($\|\mathbf{v}_\infty\|$) at departure must be minimum.

Mathematically we can express these three conditions as

$$y_1 = f_1(t_1, t_2, t_3) = \Delta v_T = 0$$

$$y_2 = f_2(t_1, t_2, t_3) = r_{\pi T}$$

$$y_3 = f_3(t_1, t_2, t_3) = \min \left\| \mathbf{v}_D \right\|,$$

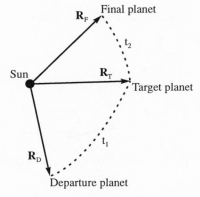

Figure 7.4   Illustration of the gravity turn.

where

$$t_1 = T_T - T_D$$

$$t_2 = T_F - T_T$$

$$t_3 = T_D - T_0.$$

Also,

- T_0 is a fixed reference date.
- Δv_T is the impulsive velocity magnitude at the target planet, which we specify to be zero.
- $r_{\pi T}$ is the specified radius of closest approach to the target planet.

The three conditions above must be solved simultaneously for the three times t_1, t_2, t_3 from which the three dates $T_D, T_T,$ and T_F are found. The solution must be found by a numerical, iterative procedure. Initial guesses for $T_D, T_T,$ and T_F are made and then systematically corrected until the three conditions are satisfied. During each iteration we use Algorithm No. 7 twice: once for the conic from the departure planet to the target planet and again for the conic from the target to the final planet. The following $\mathbf{v}_\infty$ vectors (refer to fig. 7.5) are obtained from this procedure:

$$\mathbf{v}_D = \mathbf{V}_D - \mathbf{U}_D$$

$$\mathbf{v}_{AT} = \mathbf{V}_{AT} - \mathbf{U}_T$$

$$\mathbf{v}_{DT} = \mathbf{V}_{DT} - \mathbf{U}_T$$

$$\mathbf{v}_F = \mathbf{V}_F - \mathbf{U}_F,$$

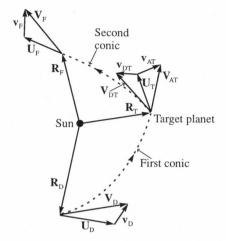

Figure 7.5   Positions and velocities of the
spacecraft for flyby of target planet in
heliocentric system.

where the subscript notation is D for departure, F for final, AT for arrival at
target planet, and DT for departure from target planet.

The *first condition* is computed from

$$y_1 = \Delta v_T = \|\mathbf{v}_{AT}\| - \|\mathbf{v}_{DT}\| .$$

When this condition is satisfied, a continuous planetocentric hyperbolic trajectory exists since the magnitudes of the arrival and departure v-infinities are the
same.

The *second condition* is

$$y_2 = r_{\pi T} .$$

When this condition is satisfied, $r_{\pi T}$ equals the magnitude of the closest approach radius vector to the target planet ($\mathbf{r}_{\pi T}$) we specify. Since the limiting
true anomaly (ϕ_L) is related to the angle θ, the radius $r_{\pi T}$ can be computed as
follows (we assume the first condition has been satisfied). In figure 7.6 we see
that the angle θ between $\mathbf{v}_{DT}$ and minus $\mathbf{v}_{AT}$ is found from the cross product,

$$\theta = \pi - \sin^{-1}\left(\frac{\|\mathbf{v}_{AT} \times \mathbf{v}_{DT}\|}{v_{AT}\, v_{DT}} \right).$$

Also in figure (7.6),

$$\phi_L + \frac{\theta}{2} = \pi.$$

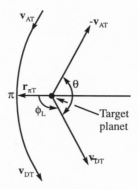

Figure 7.6 The flyby of
the target planet in the
planetocentric system.

Therefore,

$$\phi_L = \frac{\pi}{2} + \frac{1}{2} \sin^{-1} \left(\frac{\| \mathbf{v}_{AT} \times \mathbf{v}_{DT} \|}{v_{AT} \, v_{DT}} \right).$$

From §7.1.3 we can now compute the eccentricity and semi-major axis of the hyperbola,

$$e = -\frac{1}{\cos \phi_L}$$

$$a = -\frac{\mu}{v_{DT}^2} = -\frac{\mu}{v_{AT}^2},$$

where μ is the gravitational constant of the target planet; and finally,

$$r_{\pi T} = a(1 - e).$$

The *third condition* is found from

$$y_3 = \| \mathbf{v}_D \|,$$

which is the magnitude of the departure $\mathbf{v}_\infty$ vector. This condition is met when $\| \mathbf{v}_D \|$ has reached its minimum value.

We must emphasize that none of these three conditions is automatically satisfied until the correct set of dates T_D, T_T, and T_F has been found. We also note that a solution might not exist at all. A multivariable search program such as in reference [26] is essential.

It is interesting to note that this procedure can be generalized for any number of planets. For the *single* planet flyby, the "size" of the system is 3 by 3: 3

planets, 3 conditions (y_1, y_2, y_3) to be met and 3 times (t_1, t_2, t_3) to be solved for. For each additional planet inserted in the sequence we add one more condition of the type

$$y_k = \Delta v_k = 0,$$

where the subscript k refers to the kth flyby planet.

For each additional planet we also must solve for one more encounter date T_k. However, we *cannot* add another condition of the type

$$r_{\pi k} = \text{specified},$$

since this would give us more equations than unknowns.

For the N-planet sequence we will flyby $N - 2$ planets. The system of equations of conditions becomes

$$y_1 = f_1(t_1, \ldots, t_N) = \Delta v_2 = 0$$
$$y_2 = f_2(t_1, \ldots, t_N) = \Delta v_3 = 0$$
$$y_3 = f_3(t_1, \ldots, t_N) = \Delta v_4 = 0$$
$$\vdots$$
$$y_{N-2} = f_{N-2}(t_1, \ldots, t_N) = \Delta v_{N-1} = 0$$
$$y_{N-1} = f_{N-1}(t_1, \ldots, t_N) = r_{\pi k}$$
$$y_N = f_N(t_1, \ldots, t_N) = \min \|\mathbf{v}_{D1}\|,$$

where equations y_1 through y_{N-2} are the $N - 2$ free flyby conditions, $r_{\pi k}$ specifies the pericenter radius at the kth planet ($2 \leq k < N - 1$), and $\|\mathbf{v}_{D1}\|$ is the magnitude of the departure velocity at the first planet. Also,

$$t_1 = T_2 - T_1$$
$$t_2 = T_3 - T_2$$
$$t_3 = T_4 - T_3$$
$$\vdots$$
$$t_{N-1} = T_N - T_{N-1}$$
$$t_N = T_1 - T_0,$$

with $T_1 = T_{D1}$.

For an example of this technique applied to the four-planet sequence of the Earth, Jupiter, Saturn, and return to Earth, consult the paper by Bond and Anson [9].

7.2 Space Shuttle Ascent Targets

Space shuttle powered flight lasts on the order of $8\frac{1}{2}$ minutes, terminating at main engine cutoff (MECO) (fig. 7.7). During this phase of flight the twin solid rocket boosters (SRBs) and shuttle main engines (SSMEs) are used to insert the orbiter into an orbit, with the apogee altitude being the mission target altitude. (*Note*: This applies *only* to Direct Insertion [DI] missions.) For example, several shuttle missions have been designed for an operational orbit of 160 nautical miles (nm) circular. Therefore, the target orbit apogee at MECO is 160 nm.[1]

The orientation of the shuttle's orbit at MECO is designed (among other things) to allow for disposal of the External Tank (ET) in the Pacific Ocean. This requirement to a large degree bounds the true anomaly of the shuttle orbit at MECO. A typical value for the true anomaly is about 50°. Although this true anomaly places perigee over the western United States (for a 28.45° inclination mission), atmospheric drag on the ET pulls the tank into the atmosphere much sooner, usually in an area between the Marshall Islands and Hawaii.

The problem we wish to solve is the following: Given the target apogee, MECO altitude, and true anomaly at MECO, calculate the MECO velocity and flight path angle. The velocity and flight path angle are used by shuttle onboard second stage guidance software to steer the orbiter during powered flight to the desired orbit.

Note that the flight path angle (γ) in this problem is the angle between the local vertical-local horizontal (LVLH) plane and the velocity vector. The LVLH plane is at right angle to the radius vector.

We begin by first defining the position vector at apogee. Since we are provided the altitude, the apogee position vector $\mathbf{r}_a$ in the $\hat{\mathbf{P}}, \hat{\mathbf{Q}}, \hat{\mathbf{c}}$ coordinate system is

$$\mathbf{r}_a = -r_a\hat{\mathbf{P}}.$$

---

[1] Actually, the apogee MECO target is about 4 nm lower than the final desired altitude. About 18 seconds after MECO, the shuttle separates from the External Tank. Shortly after separation the remaining SSME propellant in the fuel lines is vented, providing a small additional velocity (on the order of 10 ft/sec). This extra velocity raises the apogee to the final desired altitude.

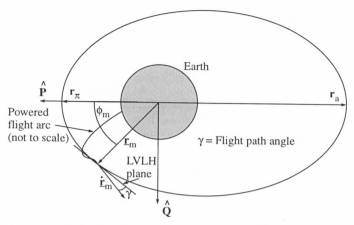

Figure 7.7   Shuttle orbit at main engine cut-off (MECO).

The position vector at MECO is, from equation (4.6),

$$\mathbf{r}_m = r_m \cos \phi_m \, \hat{\mathbf{P}} + r_m \sin \phi_m \, \hat{\mathbf{Q}}.$$

At apogee we know the local radial component of the velocity vector is zero. In other words,

$$\dot{\mathbf{r}}_a = -v_a \hat{\mathbf{Q}}.$$

Now recall equation (6.31), which we rewrite as

$$\dot{r} = \frac{1}{g} \left(\dot{g}r - r_0 \cos \Delta\phi \right).$$

Since $\dot{r}_a$ is zero, this results in

$$\dot{g}r_a = r_m \cos \Delta\phi.$$

The delta true anomaly is the angle between $\mathbf{r}_m$ and $\mathbf{r}_a$,

$$\Delta\phi = \phi_a - \phi_m = (180°) - \phi_m.$$

Recall the equation for $\dot{g}$,

$$\dot{g} = 1 - \frac{r_m}{p} \left(1 - \cos \Delta\phi \right).$$

Solve for the semi-latus rectum to obtain

$$p = \frac{r_a r_m (1 - \cos \Delta\phi)}{r_a - r_m \cos \Delta\phi}.$$

From equation (6.27) we have

$$p = \frac{r_a^2 v_a^2}{\mu},$$

where v_a is the horizontal velocity. Rearrange the above equation to get

$$v_a = \frac{\sqrt{\mu p}}{r_a}.$$

Since we now have the apogee position and velocity, we can solve for the necessary Keplerian elements to calculate the MECO velocity. The elements we need are the energy, semi-major axis, and eccentricity:

$$h = \frac{1}{2} v_a^2 - \frac{\mu}{r_a}$$

$$a = \frac{-\mu}{2h}$$

$$e = \frac{r_a}{a} - 1.$$

Now use equation (4.9),

$$\dot{\mathbf{r}}_m = \sqrt{\frac{\mu}{p}} \left[-\sin\phi_m \, \hat{\mathbf{P}} + (e + \cos\phi_m) \, \hat{\mathbf{Q}} \right],$$

for the velocity vector at MECO. The MECO flight path angle, as can be seen from figure 7.7, is

$$\gamma = 90° - \cos^{-1}\left(\frac{\mathbf{r}_m \cdot \dot{\mathbf{r}}_m}{\|\mathbf{r}_m\| \, \|\dot{\mathbf{r}}_m\|} \right)$$

or

$$\gamma = \sin^{-1}\left(\frac{\mathbf{r}_m \cdot \dot{\mathbf{r}}_m}{\|\mathbf{r}_m\| \, \|\dot{\mathbf{r}}_m\|} \right).$$

To determine the actual velocity and γ values used for shuttle missions, we must factor in several perturbations, restrictions, and constraints. However, the two-body problem does provide approximate answers usable as an initial estimate.

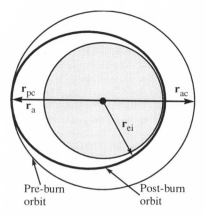

Figure 7.8   Pre- and post-burn orbits
(not to scale).

7.3 Two-Body Anytime-Deorbit Solution

In this section we detail a two-body analytic solution for the required change in
velocity (ΔV) to deorbit from a near-circular orbit. Since space shuttle orbits
are nearly circular, this solution can be used to obtain a rough idea of the ΔV
needed to deorbit. The solution to this problem relies on the assumption of a
near-circular orbit. This allows us to choose the deorbit burn point at perigee
(or apogee; in either case the flight path angle is zero), which then yields a
single solution. A deorbit burn from a point other than perigee or apogee yields
two solutions. For near-circular orbits the flight path angle is small (less than
$2°$), the cosine of which differs from 1 by at most 0.06%.

The notation we use for the derivation is as follows (see fig. 7.8):

c_1, c_2 The LTVCON linear targeting constraint parameters
r_{ac} The magnitude of the pre-burn ($\approx$ circular) orbit apogee
r_{pc} The magnitude of the pre-burn orbit perigee
r_a The magnitude of the post-burn orbit apogee
r_p The magnitude of the post-burn orbit perigee
μ The gravitational parameter ($= G(M + m)$)
r_{ei} The magnitude of the position vector at entry interface.

Throughout the derivation we assume the burn is executed at perigee. This
solution will yield the more conservative solution. The post-deorbit-burn orbit

has a perigee much lower than the pre-burn apogee or perigee. The altitude difference is larger (and requires more ΔV, i.e., more fuel) if the burn occurs at the pre-burn perigee. Entry interface is the point where the vehicle enters the atmosphere. The flight path angle at entry interface (LTVCON parameter c_2) is constrained by protection against skipping off the atmosphere (shallow angle constraint) and by stress load and frictional heating limits (steep angle constraint). The LTVCON constraint on the velocity is

$$v_r = c_1 + c_2 v_h, \qquad (7.1)$$

where v_r is the radial component of the velocity vector and v_h is the horizontal component.

We first derive the equation for the energy integral of the motion for the post-burn orbit at r_{ei} in terms of the orbit's apogee (r_a) and perigee (r_p). Observe that the post-burn apogee is given by

$$r_a = r_{pc} \ (\simeq r_{ac}).$$

The energy integral is

$$h = \frac{1}{2} \mathbf{v}_{ei} \cdot \mathbf{v}_{ei} - \frac{\mu}{r_{ei}}, \qquad (7.2)$$

where $\mathbf{v}_{ei}$ is the velocity vector at entry interface, given by

$$\mathbf{v}_{ei} = v_r \, \hat{\mathbf{r}} + v_h \, \hat{\mathbf{h}}$$
$$= (c_1 + c_2 v_h) \hat{\mathbf{r}} + v_h \, \hat{\mathbf{h}}. \qquad (7.3)$$

The $\hat{\mathbf{r}}, \hat{\mathbf{h}}, \hat{\mathbf{c}}$ coordinate system is illustrated in figure 7.9. We can relate the energy to the semi-major axis using

$$h = -\frac{\mu}{2a} = -\frac{\mu}{r_a + r_p}. \qquad (7.4)$$

The dot product of the velocity vector $\mathbf{v}_{ei}$ is

$$\mathbf{v}_{ei} \cdot \mathbf{v}_{ei} = (c_1 + c_2 v_h)^2 + v_h{}^2. \qquad (7.5)$$

Substituting equations (7.4) and (7.5) into equation (7.2) and solving for the squared velocity gives

$$2\mu \left(\frac{r_a + r_p - r_{ei}}{r_{ei}(r_a + r_p)} \right) = (c_1 + c_2 v_h)^2 + v_h{}^2. \qquad (7.6)$$

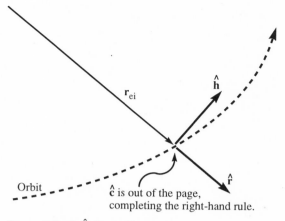

Figure 7.9 $\hat{\mathbf{r}}, \hat{\mathbf{h}}, \hat{\mathbf{c}}$ coordinate system.

We need an expression for the post-burn orbit perigee, r_p. From the equation of a conic,

$$r_p = \frac{p}{1 + e \cos 0°} = \frac{p}{1 + e}, \tag{7.7}$$

since the perigee true anomaly is zero. Equations for the semi-latus rectum and eccentricity are (from chapter 3),

$$p = \frac{c^2}{\mu} \tag{7.8}$$

$$e = \frac{r_a - r_p}{r_a + r_p}. \tag{7.9}$$

Now calculate the area integral by taking the cross product of the radius and velocity vectors expressed in the $\hat{\mathbf{r}}, \hat{\mathbf{h}}, \hat{\mathbf{c}}$ coordinate system:

$$\begin{aligned}
\mathbf{c} &= \mathbf{r}_{ei} \times \mathbf{v}_{ei} \\
&= r_{ei}\,\hat{\mathbf{r}} \times \left((c_1 + c_2 v_h)\hat{\mathbf{r}} + v_h\,\hat{\mathbf{h}} \right) \\
c\,\hat{\mathbf{c}} &= r_{ei}\, v_h \hat{\mathbf{c}} \\
c &= r_{ei}\, v_h. \tag{7.10}
\end{aligned}$$

Using equations (7.8)–(7.10) in equation (7.7) and solving for r_p results in

$$r_p = \frac{r_{ei}^2 \, v_h^2 \, r_a}{2r_a \, \mu - r_{ei}^2 \, v_h^2}. \tag{7.11}$$

Substituting equation (7.11) for r_p into equation (7.6), and after some algebra, we get

$$A v_h^2 + B v_h + C = 0,$$

where

$$A = \frac{r_a^2}{r_{ei}^2} \left(c_2^2 + 1 \right) - 1$$

$$B = 2c_1 c_2 \frac{r_a^2}{r_{ei}^2}$$

$$C = \frac{r_a^2}{r_{ei}^2} \left(c_1^2 - \frac{2\mu}{r_{ei}} \right) + \frac{2\mu r_a}{r_{ei}^2}.$$

We have, therefore,

$$v_h = \frac{-B + \sqrt{B^2 - 4AC}}{2A}.$$

Note that we choose the positive root since v_h is a positive value. Using equation (7.1) for v_r, we now have the components of $\mathbf{v}_{ei}$, the magnitude of which is

$$v_{ei} = \sqrt{v_r^2 + v_h^2}.$$

Obtaining $\mathbf{v}_{ei}$ permits calculation of the post-burn orbital energy,

$$h = \frac{1}{2} v_{ei}^2 - \frac{\mu}{r_{ei}} = \frac{1}{2} v_a^2 - \frac{\mu}{r_a},$$

where v_a is the velocity of the vehicle at apogee in the post-burn orbit. Solving for v_a, we obtain,

$$v_a = \sqrt{v_{ei}^2 - 2\mu \frac{r_a - r_{ei}}{r_a r_{ei}}}, \tag{7.12}$$

where v_a is the vehicle velocity immediately after the deorbit burn. To determine the ΔV for the burn, we need the vehicle's velocity in the pre-burn orbit at the same point (the pre-burn orbital perigee, r_{pc}). This quantity can be obtained by solving the energy integral equation in the pre-burn orbit,

$$h_{pre\text{-}burn} = \frac{1}{2}v_{pc}^2 - \frac{\mu}{r_{pc}} = -\frac{\mu}{r_{pc} + r_{ac}}.$$

Solving for the pre-burn orbital velocity at perigee (v_{pc}) yields

$$v_{pc} = \sqrt{\frac{2\mu r_{ac}}{r_{pc}(r_{pc} + r_{ac})}}. \tag{7.13}$$

The value for the required ΔV is now given by

$$\Delta V = v_{pc} - v_a. \tag{7.14}$$

Using equation (7.14), where v_{pc} is given by equation (7.11) and v_a is given by equation (7.12), we can calculate the necessary ΔV to deorbit from a near-circular orbit and reach entry interface, satisfying the altitude (r_{ei}) and flight path angle (c_2) constraints.

7.4 Relative Motion

In this section we discuss the motion of two satellites (m_1 and m_2) relative to each other. Each body has unit mass and is orbiting a third (massive) body such as the Earth (m_3). We can therefore calculate the orbit of each satellite using two-body methods developed in chapters 4 and 5.

We illustrate an approach to studying relative motion with four examples. To obtain the solution we will set up initial position and velocity vectors for the two orbits, propagate the orbits forward by a time step of $T_1/360$ (T_1 is the period of m_1), and plot the relative position $\mathbf{r}$ given by

$$\mathbf{r} = \mathbf{r}_2 - \mathbf{r}_1.$$

With this definition the plotted motion is m_2 relative to m_1; that is, m_2 will appear to move about a fixed m_1. For the first three examples, we plot the relative motion in both the inertial $\hat{\mathbf{P}}, \hat{\mathbf{Q}}, \hat{\mathbf{c}}$, and rotating LVLH (local vertical, local horizontal) coordinate systems of m_1. An LVLH plane is illustrated in figure 7.7 and defined in §7.2.

We first develop an expectation for the results using two-body dynamics. We then solve the problem and plot the relative motion for two orbits using *Mathematica* [77]. Students are encouraged to develop algorithms to explore the many variations possible. Understanding how spacecraft move relative to each other is a vital skill needed with the advent of space shuttle satellite repair and space station rendezvous missions.

For each example the orbit of m_1 is defined by

$$\mathbf{r}_1 = 6800.0 \, \text{km} \, \hat{\mathbf{P}}$$

$$e_1 = 0.35,$$

and we use $\mu = 398601.0$ for the gravitational constant.

Recall from chapter 2 that to define an orbit we need six integrals or constants. For our definition of m_1 these are

- Position (3 constants).
- $\hat{\mathbf{P}}$ and $\hat{\mathbf{c}}$ components of the velocity are zero.
- Eccentricity.

We will show below that since m_1 is at perigee, the $\hat{\mathbf{P}}$ and $\hat{\mathbf{c}}$ components of the velocity are zero.

7.4.1 Radial Displacement

The first example has the initial conditions,

$$\mathbf{r}_2 = 6810.46 \, \text{km} \, \hat{\mathbf{P}}$$

$$e_2 = e_1 = 0.35.$$

Note that the initial positions are both at perigee. We choose this because we know the velocity vector is then oriented along the $\hat{\mathbf{Q}}$ axis and facilitates calculating $\dot{\mathbf{r}}_1$ and $\dot{\mathbf{r}}_2$ (why?).

Since we have the direction we next calculate the magnitude v_1 of the velocity vector $\dot{\mathbf{r}}_1$ as follows. The magnitude of the angular momentum vector is

$$c = \|\mathbf{c}\| = r_1 \, v_1 \, \sin 90^\circ = r_1 \, v_1 = \sqrt{\mu p_1},$$

where we take advantage of working at perigee: $\mathbf{r}_1$ and $\dot{\mathbf{r}}_1$ are orthogonal. Solve for v_1 and use

$$p_1 = r_1(1 + e_1 \cos(0^\circ)) = r_1(1 + e_1)$$

to obtain

$$v_1 = \frac{\sqrt{\mu r_1(1 + e_1)}}{r_1}$$

$$= \sqrt{\frac{\mu(1 + e_1)}{r_1}}.$$

We prove the velocity vector direction and calculate its components by an alternate approach (use eq. 4.9) as follows. The true anomaly at perigee is zero. With this and the equation for p_1,

$$
\begin{aligned}
\dot{\mathbf{r}}_1 &= \sqrt{\frac{\mu}{p_1}} \left[-\sin\phi\, \hat{\mathbf{P}} + (e_1 + \cos\phi)\, \hat{\mathbf{Q}} \right] \\
&= \sqrt{\frac{\mu}{r_1(1+e_1)}} \left[-\sin 0°\, \hat{\mathbf{P}} + (e + \cos 0°)\, \hat{\mathbf{Q}} \right] \\
&= \sqrt{\frac{\mu}{r_1(1+e_1)}}(1+e_1)\hat{\mathbf{Q}} \\
&= \sqrt{\frac{\mu(1+e_1)}{r_1}}\hat{\mathbf{Q}} = 8.8957\,\text{km/sec}\,\hat{\mathbf{Q}},
\end{aligned}
$$

which is the earlier result. Similarly, the $\dot{\mathbf{r}}_2$ vector is given by

$$
\dot{\mathbf{r}}_2 = \sqrt{\frac{\mu(1+e_2)}{r_2}}\hat{\mathbf{Q}} = 8.8889\,\text{km/sec}\,\hat{\mathbf{Q}}.
$$

We demonstrate several approaches here to emphasize the advantage of choosing our initial conditions at perigee. The right choices for those aspects of a problem over which we have control can greatly simplify the problem.

First we develop an expectation for the difference in orbital periods. Since the period is proportional to the semi-major axis (recall from Kepler's third law that $T^2 \propto a^3$), and the two orbits have different semi-major axes, we would expect that the relative separation increases with time. With m_2 having the larger semi-major axis, it has the longer period. We next develop an expectation for the orientation of the motion. When m_1 has reached the point where its true anomaly (ϕ) is 90°, the position vector $\mathbf{r}_1$ lies along the $\hat{\mathbf{Q}}$ axis and has a magnitude equal to the semi-latus rectum (easily seen from the equation of a conic). However, m_2 still has a positive $\hat{\mathbf{P}}$ component since it is lagging behind. Therefore, $\mathbf{r}$ is either in the first or fourth quadrant.

After $\frac{1}{2}$ period for m_1 the position vector $\mathbf{r}_1$ lies along $-\hat{\mathbf{P}}$. However, again m_2 lags behind and has not yet arrived at apogee and therefore has a component lying along the positive $\hat{\mathbf{Q}}$ axis. Therefore, $\mathbf{r}$ lies in either the first or second quadrant.

These two observations suggest that the relative motion has traveled up through the first quadrant and well into the second. Since the relative distance is increasing, we apparently have positively oriented spiral-like relative motion, the first two orbits of which are shown in figure 7.10.

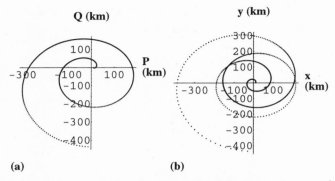

Figure 7.10   Inertial (a) and LVLH (b) relative motion for example 1 (radial displacement).

7.4.2 Varying Eccentricity

For the second example the satellite m_2 has the initial conditions,

$$\mathbf{r}_2 = \mathbf{r}_1 = 6800.0 \, \text{km} \, \hat{\mathbf{P}}$$

$$e_2 = 0.349.$$

The initial velocity vectors are therefore

$$\dot{\mathbf{r}}_1 = \sqrt{\frac{\mu(1 + e_1)}{r_1}} \hat{\mathbf{Q}} = 8.8957 \, \text{km/sec} \, \hat{\mathbf{Q}}$$

$$\dot{\mathbf{r}}_2 = \sqrt{\frac{\mu(1 + e_2)}{r_2}} \hat{\mathbf{Q}} = 8.8924 \, \text{km/sec} \, \hat{\mathbf{Q}}.$$

Note that m_2 is again traveling slower than m_1. With the periods again not equal we expect the same spiral motion as the previous example. This is indeed what we find as illustrated in figure 7.11. Note that if we instead increase e_2 to $e_1 + \Delta e$, then m_2 would have a larger initial velocity. We leave this case for students to evaluate.

7.4.3 Periodic Motion

The initial conditions for the third example combine the changes for m_2 from the previous two examples,

$$\mathbf{r}_2 = 6810.46 \, \hat{\mathbf{P}}$$

$$e_2 = 0.349.$$

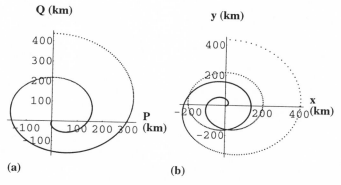

Figure 7.11   Inertial (a) and LVLH (b) relative motion for example 2 (varying the eccentricity).

We begin by calculating the periods of the orbits using Kepler's third law:

$$p_1 = r_1(1 + e_1) = 9180.0 \text{ km}$$

$$a_1 = \frac{p_1}{(1 - e_1^2)}$$

$$= \frac{r_1(1 + e_1)}{(1 + e_1)(1 - e_1)}$$

$$= \frac{r_1}{(1 - e_1)}$$

$$T_1 = \sqrt{\left(\frac{4\pi^2}{\mu}\right) a_1^3}$$

$$= \sqrt{\frac{4\pi^2 r_1^3}{\mu(1 - e_1)^3}}$$

$$= 10648.9 \text{ sec}$$

and

$$p_2 = r_2(1 + e_2) = 9187.3 \text{ km}$$

$$T_2 = \sqrt{\frac{4\pi^2 r_2^3}{\mu(1 - e_2)^3}}$$

$$= 10648.9 \text{ sec}.$$

Both the period and energy of an orbit are proportional to the semi-major axis (recall that $h = -\mu(2a)^{-1}$). Therefore, if $a_1 = a_2$ or $h_1 = h_2$, then we have

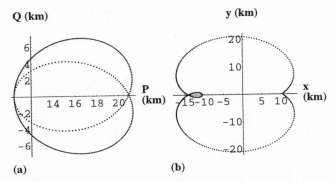

Figure 7.12   Inertial (a) and LVLH (b) periodic relative motion of example 3.

periodic motion. Both m_1 and m_2 return to perigee at the same time. The results for this example are plotted in figure 7.12.

Also, since the periods are equal, then half the periods are also equal, so both bodies arrive at apogee at the same time. When the masses have reached apogee, we have

$$\mathbf{r}_1 = -\frac{p_1}{1 - e_1}\,\hat{\mathbf{P}} = 14123.1 \text{ km}$$

$$\mathbf{r}_2 = -\frac{p_2}{1 - e_2}\,\hat{\mathbf{P}} = 14112.1 \text{ km}.$$

We see therefore that the masses have changed places: m_2 is now lower than m_1. As the orbits proceed from apogee back to perigee, the shape of the relative motion plot is reflected about the $\hat{\mathbf{P}}$ axis. This is due to *both* $\mathbf{r}_1$ and $\mathbf{r}_2$ having changed signs and relative positions. We show this in figure 7.13.

This example initialized both m_1 and m_2 at perigee. Therefore the difference in true anomalies is zero. It is important to understand that *this difference* is what is repeated twice during an orbit. If the orbits were initialized with a $\Delta\phi$ of 42°, then $\frac{1}{2}$ orbit later $\Delta\phi$ would again be 42° *and at no other time* (there is one exception—what is it?). Students can convince themselves of this by setting up this example, rotating the initial $\mathbf{r}_2$ and $\dot{\mathbf{r}}_2$ vectors through an angle $\Delta\phi$ and observing that instead of the relative motion orbits intersecting on the $\hat{\mathbf{P}}$ axis they will intersect at a point $\Delta\phi$ degrees relative to $\hat{\mathbf{P}}$ (above or below depending on the rotation matrix). Refer to Appendix A for a discussion of rotation matrices. The exception we refer to is if both orbits are circular, for then $\Delta\phi$ would be constant.

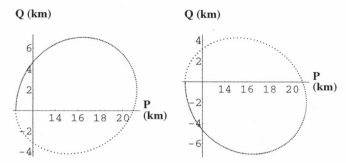

Figure 7.13 Inertial periodic relative motion plots of example 3 for 1/2 period.

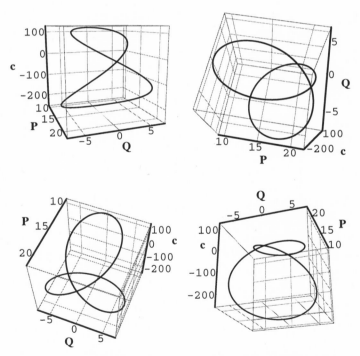

Figure 7.14 Inertial three-dimensional periodic motion of example 4.

7.4.4 *Rotation about the* $\hat{Q}$ *Axis*

For the fourth example we add a rotation about the $\hat{Q}$ axis to introduce three-dimensional relative motion. The initial conditions for m_2 are

$$e_2 = 0.349$$

$$\mathbf{r}_2 = \begin{pmatrix} \cos 1° & 0 & -\sin 1° \\ 0 & 1 & 0 \\ \sin 1° & 0 & \cos 1° \end{pmatrix} \begin{pmatrix} 6810.46 \\ 0 \\ 0 \end{pmatrix}$$

$$\dot{\mathbf{r}}_2 = \begin{pmatrix} \cos 1° & 0 & -\sin 1° \\ 0 & 1 & 0 \\ \sin 1° & 0 & \cos 1° \end{pmatrix} \begin{pmatrix} 0 \\ \sqrt{\mu(1+e_2)/r_2} \\ 0 \end{pmatrix},$$

where the initialization of $\mathbf{r}_2$ is the initial position of m_2 from the previous two examples rotated $1°$ about the $\hat{Q}$ axis.

Observe that at perigee $\mathbf{r}_2$ has a small component lying in the positive $\hat{c}$ direction but $\dot{\mathbf{r}}_2$ is still oriented along $\hat{Q}$. What we find at apogee is the reverse: $\dot{\mathbf{r}}_2$ lies along $-\hat{Q}$ and $\mathbf{r}_2$ has a small component lying along the *negative* $\hat{c}$ axis. Recall that m_1 and m_2 arrive at perigee and apogee at the same time. With m_2 above the $\hat{P}, \hat{Q}$ plane for half an orbit and below the plane for the other half, we would expect the relative motion to be split above and below the $\hat{P}, \hat{Q}$ plane. Figure 7.14 shows four different viewpoints of the three-dimensional motion.

II

PERTURBATION METHODS

8

PERTURBATION THEORY

8.1 Introduction

In this chapter we begin the study of perturbation methods. We saw in chapter 2 that the two-body problem is a completely solvable problem since it has the required number (six) of integrals, or constants, of the motion. When accelerations other than mutual gravitational attraction between the two masses occur, the solution for the six integrals may no longer be possible. In general, the integrals are no longer constant but become functions of time. However, depending on the type of non-two-body acceleration imposed, some of the integrals might remain constant, or some new integrals might appear. It is certain that the imposed non-two-body acceleration, which we call a *perturbation,* will render the new system insolvable; that is, the perturbation will always reduce the number of integrals.

The temptation at this point is to accept our misfortune and simply integrate the system of differential equations by a numerical technique. This unsophisticated approach throws away all of our knowledge of the two-body problem and its integrals. The solution of the perturbed two-body problem in cartesian coordinates,

$$\ddot{\mathbf{r}} + \frac{\mu}{r^3}\mathbf{r} = \mathbf{F},$$

by a numerical method is called *Cowell's method.* In this method the perturbation ($\mathbf{F}$) is treated the same as the two-body part ($\frac{\mu}{r^3}\mathbf{r}$). The advantage of this method is simplicity. The disadvantage of Cowell's method is that small errors introduced at any stage cause the numerical solution to diverge from the true solution very quickly [15][21]. The presence of the perturbation contributes to the divergence of the numerical solution.

Another approach to the solution of the perturbed two-body problem by a numerical method is known as *Encke's method.* This method solves for the deviation (η) of the solution (**r**) of the perturbed problem from the solution (ρ) of the unperturbed problem,

$$\ddot{\rho} + \frac{\mu}{\rho^3}\rho = 0.$$

Subtracting these two differential equations results in the differential equation for the deviations,

$$\ddot{\eta} = \frac{\mu}{\rho^3}\left[\left(1 - \frac{\rho^3}{r^3}\right)(\eta + \rho) - \eta\right] + \mathbf{F},$$

where

$$\eta = \mathbf{r} - \rho,$$

and the initial conditions are $\eta_0 = 0$ and $\dot{\eta}_0 = 0$. The advantage of Encke's method over Cowell's method is that the deviations are smaller in magnitude than either solution (**r** or ρ). However, the error growth is the same as that for Cowell's method—small errors cause the numerical solution to diverge [15]. An additional disadvantage is that the solution process must be restarted (a step called *rectification*) when the magnitude of the deviations becomes large.

A more intelligent approach is to consider the integrals of the motion as new dependent variables for which we then develop a new system of differential equations. This approach has the advantage of (1) giving us insight on the effect of the perturbation on the integrals of the motion; (2) at times presenting us with the opportunity of solving the new differential equations by some approximations; and (3) giving us a new system of differential equations that are more amenable to numerical solution since the integrals in many cases are almost constant. The development of this new system of differential equations is the essence of perturbation methods or, as they are sometimes called, *variation of parameters* methods. In this section we will develop and apply two of these methods: Poisson's method and Lagrange's method. But first we give a short explanation of the method of variation of parameters and illustrate with a familiar example.

8.1.1 Explanation of Perturbation Theory

Suppose we have a *first order system*

$$\dot{\mathbf{x}} = \mathbf{f}(\mathbf{x}, t), \tag{8.1}$$

where

$$\mathbf{x}^T = (x_1, x_2, \ldots, x_n) \text{ and } \mathbf{f}^T = (f_1, f_2, \ldots, f_n),$$

which has the solution

$$\mathbf{x} = \mathbf{x}(\mathbf{c}, t), \qquad (8.2)$$

where

$$\mathbf{c}^T = (c_1, c_2, \ldots, c_n)$$

are the constants of integration (or integrals of the motion). Note that the system (8.2) is completely solvable.

Now let the system (8.1) be modified by a function

$$\mathbf{g}^T(\mathbf{x}, t) = (g_1, g_2, \ldots, g_n),$$

which is called a perturbation. The *perturbed* system is

$$\dot{\mathbf{x}} = \mathbf{f}(\mathbf{x}, t) + \mathbf{g}(\mathbf{x}, t). \qquad (8.3)$$

We now modify the solution (eq. 8.2) such that it has the form

$$\mathbf{x} = \mathbf{x}(\mathbf{c}(t), t). \qquad (8.4)$$

Observe that the solutions (8.2) and (8.4) are the same except that the components of $\mathbf{c}$ are now functions of the independent variable. The objective now is to find differential equations for the functions $c_k(t)$.

8.1.2 Elementary Example—Harmonic Oscillator

The differential equation for the *unperturbed* system is

$$\ddot{x} + x = 0, \qquad (8.5)$$

which has the solution

$$x = a \cos t + b \sin t \qquad (8.6)$$

$$\dot{x} = -a \sin t + b \cos t. \qquad (8.7)$$

The initial conditions (at $t = 0$) are the constants a and b,

$$x(0) = a \quad \text{and} \quad \dot{x}(0) = b. \tag{8.8}$$

The differential equation for the *perturbed* system is

$$\ddot{x} + x = g(x, t). \tag{8.9}$$

The solution of equation (8.9) has the same form as equations (8.6) and (8.7), but a and b are now functions of time. We develop the solution to equation (8.9) by differentiating equation (8.6) with $a = a(t)$ and $b = b(t)$,

$$\dot{x} = -a \sin t + b \cos t + \dot{a} \cos t + \dot{b} \sin t. \tag{8.10}$$

The solution procedure requires that equation (8.10) have the same form as equation (8.7); therefore,

$$\dot{a} \cos t + \dot{b} \sin t = 0. \tag{8.11}$$

Differentiating equation (8.7) with $a = a(t)$ and $b = b(t)$, we obtain

$$\ddot{x} = -a \cos t - b \sin t - \dot{a} \sin t + \dot{b} \cos t. \tag{8.12}$$

Substitute equation (8.6) into equation (8.12) to get

$$\ddot{x} = -x - \dot{a} \sin t + \dot{b} \cos t. \tag{8.13}$$

Comparing equation (8.13) with equation (8.9) (the perturbed differential equation), we see that

$$-\dot{a} \sin t + \dot{b} \cos t = g. \tag{8.14}$$

We now have two equations (8.11 and 8.14) that can be solved for the unknowns $\dot{a}$ and $\dot{b}$. After some algebra we arrive at

$$\dot{a} = -g \sin t \tag{8.15}$$

$$\dot{b} = g \cos t. \tag{8.16}$$

Note that no assumptions as to the form of the perturbation $g(x, t)$ have been made. Equations (8.6) and (8.7) provide the solution to equation (8.9) with $a = a(t)$ and $b = b(t)$, found by solving the differential equations (8.15) and (8.16). This solution in turn depends on the form of the perturbation $g(x, t)$.

8.2 Poisson's Method

The derivation of this method is classical, but we use Giacaglia [33] as a reference. Consider the completely solvable second-order system

$$\ddot{\mathbf{x}} = \mathbf{f}(\mathbf{x}, \dot{\mathbf{x}}, t),\qquad (8.17)$$

where

$$\mathbf{x}^T = (x_1, x_2, \ldots, x_m)\quad \text{and}\quad \mathbf{f}^T = (f_1, f_2, \ldots, f_m).$$

Let the $2m$ integrals of the motion be given by

$$\boldsymbol{\sigma} = \boldsymbol{\sigma}(\mathbf{x}, \dot{\mathbf{x}}, t) = constant,\qquad (8.18)$$

where

$$\boldsymbol{\sigma}^T = (\sigma_1, \sigma_2, \ldots, \sigma_{2m}).$$

Observe that for any σ_k $(1 \leq k \leq 2m)$, we have along any solution $\mathbf{x}(t)$ of equation (8.17) the equation

$$\dot{\sigma}_k = \frac{\partial \sigma_k}{\partial \mathbf{x}}\dot{\mathbf{x}} + \frac{\partial \sigma_k}{\partial \dot{\mathbf{x}}}\ddot{\mathbf{x}} + \frac{\partial \sigma_k}{\partial t} = 0,$$

since σ_k is a constant. Note that the partials with respect to $\mathbf{x}$ and $\dot{\mathbf{x}}$ are each $1 \times m$ row matrices. Using equation (8.17),

$$\dot{\sigma}_k = \frac{\partial \sigma_k}{\partial \mathbf{x}}\dot{\mathbf{x}} + \frac{\partial \sigma_k}{\partial \dot{\mathbf{x}}}\mathbf{f} + \frac{\partial \sigma_k}{\partial t} = 0.$$

Now consider the *perturbed* system

$$\ddot{\mathbf{x}} = \mathbf{f}(\mathbf{x}, \dot{\mathbf{x}}, t) + \mathbf{g}(\mathbf{x}, \dot{\mathbf{x}}, t).\qquad (8.19)$$

The total derivative of σ_k along any solution of equation (8.19) is

$$\dot{\sigma}_k = \frac{\partial \sigma_k}{\partial \mathbf{x}}\dot{\mathbf{x}} + \frac{\partial \sigma_k}{\partial \dot{\mathbf{x}}}\ddot{\mathbf{x}} + \frac{\partial \sigma_k}{\partial t} \neq 0,$$

since σ_k is no longer constant. Using equation (8.19) this becomes

$$\dot{\sigma}_k = \frac{\partial \sigma_k}{\partial \mathbf{x}}\dot{\mathbf{x}} + \frac{\partial \sigma_k}{\partial \dot{\mathbf{x}}}(\mathbf{f} + \mathbf{g}) + \frac{\partial \sigma_k}{\partial t}.\qquad (8.20)$$

Now introduce the *condition* that

$$\dot{\sigma}_k = \frac{\partial \sigma_k}{\partial \dot{\mathbf{x}}} \mathbf{g};$$ (8.21)

then equation (8.20) becomes

$$\frac{\partial \sigma_k}{\partial \mathbf{x}} \dot{\mathbf{x}} + \frac{\partial \sigma_k}{\partial \dot{\mathbf{x}}} \mathbf{f} + \frac{\partial \sigma_k}{\partial t} = 0,$$ (8.22)

which is the same as in the unperturbed case.

8.2.1 Elementary Example—Harmonic Oscillator

As before, the perturbed system is

$$\ddot{x} = -x + g(x, \dot{x}, t),$$

where $m = 1$, since we are considering only a single dimension.

The solution for the unperturbed case ($g = 0$) is

$$x = a \cos t + b \sin t$$

$$\dot{x} = -a \sin t + b \cos t,$$

where a and b are constants, or integrals of the motion. Solve for the integrals $a(= \sigma_1)$ and $b(= \sigma_2)$,

$$\sigma_1 = a = x \cos t - \dot{x} \sin t$$ (8.23)

$$\sigma_2 = b = x \sin t + \dot{x} \cos t.$$ (8.24)

For this example, equation (8.21) becomes

$$\begin{pmatrix} \dot{a} \\ \dot{b} \end{pmatrix} = \begin{pmatrix} \frac{\partial a}{\partial \dot{x}} \\ \frac{\partial b}{\partial \dot{x}} \end{pmatrix} g.$$

Taking the partial derivative of equations (8.23) and (8.24), we obtain

$$\frac{\partial a}{\partial \dot{x}} = -\sin t$$

$$\frac{\partial b}{\partial \dot{x}} = \cos t.$$

Therefore,

$$\dot{a} = -g \sin t$$
$$\dot{b} = g \cos t,$$

which is the same solution we obtained in §8.1.2.

8.3 Lagrange Variation of Parameters

The derivation of the Lagrange Variation of Parameters (VOP) method is also classical, but we use Pollard [58] as a reference. We begin again with the system

$$\dot{\mathbf{x}} = \mathbf{f}(\mathbf{x}, t), \tag{8.25}$$

where

$$\mathbf{x}^T = (x_1, x_2, \ldots, x_n),$$

which has the solution

$$\mathbf{x} = \mathbf{x}(\mathbf{c}, t), \tag{8.26}$$

where

$$\mathbf{c}^T = (c_1, c_2, \ldots, c_n),$$

and again introduce a perturbation

$$\mathbf{g}(\mathbf{x}, t)^T = (g_1, g_2, \ldots, g_n),$$

which results in

$$\dot{\mathbf{x}} = \mathbf{f}(\mathbf{x}, t) + \mathbf{g}(\mathbf{x}, t). \tag{8.27}$$

As before, our goal is to make the solution of the unperturbed system valid for the perturbed system by modifying the solution (eq. 8.26) such that it has the same form but with $\mathbf{c}$ now a function of time,

$$\mathbf{x} = \mathbf{x}(\mathbf{c}(t), t). \tag{8.28}$$

Now take the total derivative of equation (8.28),

$$\dot{\mathbf{x}} = \frac{\partial \mathbf{x}}{\partial \mathbf{c}} \dot{\mathbf{c}} + \frac{\partial \mathbf{x}}{\partial t}. \tag{8.29}$$

Taking the total derivative of the unperturbed solution, equation (8.26), where we note that $x = x(t)$,

$$\dot{\mathbf{x}} = \frac{\partial \mathbf{x}}{\partial t} = \mathbf{f}, \tag{8.30}$$

and where we have used equation (8.25). Substitute equation (8.30) into equation (8.29) to get

$$\dot{\mathbf{x}} = \frac{\partial \mathbf{x}}{\partial \mathbf{c}} \dot{\mathbf{c}} + \mathbf{f}. \tag{8.31}$$

Compare equation (8.31) with equation (8.27) to obtain

$$\frac{\partial \mathbf{x}}{\partial \mathbf{c}} \dot{\mathbf{c}} = \mathbf{g}. \tag{8.32}$$

Notice that $\frac{\partial \mathbf{x}}{\partial \mathbf{c}}$ is a matrix

$$\frac{\partial \mathbf{x}}{\partial \mathbf{c}} = \begin{pmatrix} \frac{\partial x_1}{\partial c_1} & \frac{\partial x_1}{\partial c_2} & \cdots & \frac{\partial x_1}{\partial c_n} \\ \frac{\partial x_2}{\partial c_1} & \frac{\partial x_2}{\partial c_2} & \cdots & \frac{\partial x_2}{\partial c_n} \\ \vdots & \vdots & \ddots & \vdots \\ \frac{\partial x_n}{\partial c_1} & \frac{\partial x_n}{\partial c_2} & \cdots & \frac{\partial x_n}{\partial c_n} \end{pmatrix},$$

and we require this matrix to be invertible since we wish to solve equation (8.32) for $\dot{\mathbf{c}}$; therefore,

$$DET \left[\frac{\partial \mathbf{x}}{\partial \mathbf{c}} \right] \neq 0.$$

The partials are obtained from equation (8.26). The system of differential equations for the parameter $\mathbf{c}$ is obtained from equation (8.32):

$$\dot{\mathbf{c}} = \left[\frac{\partial \mathbf{x}}{\partial \mathbf{c}} \right]^{-1} \mathbf{g}. \tag{8.33}$$

8.3.1 Elementary Example—Harmonic Oscillator

The perturbed differential equation is

$$\ddot{x} = -x + g.$$

Let $x_1 = x$ and $x_2 = \dot{x}$; then

$$\dot{x}_1 = \dot{x} = x_2$$
$$\dot{x}_2 = \ddot{x} = -x_1 + g$$

or

$$\begin{pmatrix} \dot{x}_1 \\ \dot{x}_2 \end{pmatrix} = \begin{pmatrix} x_2 \\ -x_1 \end{pmatrix} + \begin{pmatrix} 0 \\ g \end{pmatrix},$$

which is in the form

$$\dot{\mathbf{x}} = \mathbf{f} + \mathbf{g}.$$

When $\mathbf{g} = 0$ the solution is

$$x_1 = a \cos t + b \sin t$$
$$x_2 = -a \sin t + b \cos t,$$

where the integration constants are

$$\mathbf{c} = \begin{pmatrix} a \\ b \end{pmatrix}.$$

When $\mathbf{g} \neq 0$, the differential equation for $\mathbf{c}$ (from equation (8.32)) is

$$\frac{\partial \mathbf{x}}{\partial \mathbf{c}} \dot{\mathbf{c}} = \mathbf{g}.$$

Thus we have

$$\begin{pmatrix} \cos t & \sin t \\ -\sin t & \cos t \end{pmatrix} \begin{pmatrix} \dot{a} \\ \dot{b} \end{pmatrix} = \begin{pmatrix} 0 \\ g \end{pmatrix}.$$

Since the partials matrix is orthogonal, multiply both sides by the transpose,

$$\begin{pmatrix} \cos t & -\sin t \\ \sin t & \cos t \end{pmatrix} \begin{pmatrix} \cos t & \sin t \\ -\sin t & \cos t \end{pmatrix} \begin{pmatrix} \dot{a} \\ \dot{b} \end{pmatrix} = \begin{pmatrix} \cos t & -\sin t \\ \sin t & \cos t \end{pmatrix} \begin{pmatrix} 0 \\ g \end{pmatrix},$$

which reduces to

$$\begin{pmatrix} 1 & 0 \\ 0 & 1 \end{pmatrix} \begin{pmatrix} \dot{a} \\ \dot{b} \end{pmatrix} = \begin{pmatrix} -g \sin t \\ g \cos t \end{pmatrix}.$$

Therefore, as before, we obtain

$$\dot{a} = -g \sin t$$
$$\dot{b} = g \cos t.$$

8.4 Two-Body Integrals of Motion

We summarize for reference the integrals of motion for the two-body problem below, which were derived in chapter 2. Note a change of notation for the Laplace vector from $\mathbf{P}$ to $\mu\boldsymbol{\varepsilon}$, where $\boldsymbol{\varepsilon}$ is the eccentricity vector.

The *fundamental* integrals were developed by vector manipulation of

$$\ddot{\mathbf{r}} + \frac{\mu}{r^3}\mathbf{r} = 0.$$

1. The *area integral* (angular momentum),

$$\mathbf{c} = \mathbf{r} \times \dot{\mathbf{r}} = constant.$$

2. The *energy integral*,

$$h = \frac{1}{2}\dot{\mathbf{r}} \cdot \dot{\mathbf{r}} - \frac{\mu}{r} = constant.$$

3. The *Laplacian integral* (eccentricity vector),

$$\mathbf{P} = \mu\boldsymbol{\varepsilon} = -\frac{\mu}{r}\mathbf{r} - \mathbf{c} \times \dot{\mathbf{r}} = constant.$$

4. The *time of pericenter passage*,

$$t_\pi = t - \sqrt{\frac{a^3}{\mu}}\,(E - e\sin E) = constant,$$

where

$$r = a(1 - e\cos E)$$
$$a = \frac{-\mu}{2h}$$
$$e = \|\boldsymbol{\varepsilon}\|.$$

Note that these 8 quantities ($\mathbf{c}, \boldsymbol{\varepsilon}, h, t_\pi$) are not independent since we have the two relations,

$$\mathbf{c} \cdot \boldsymbol{\varepsilon} = 0$$

and

$$p \equiv a(1 - e^2) = \frac{c^2}{\mu} = \frac{-\mu}{2h}(1 - \boldsymbol{\varepsilon} \cdot \boldsymbol{\varepsilon}).$$

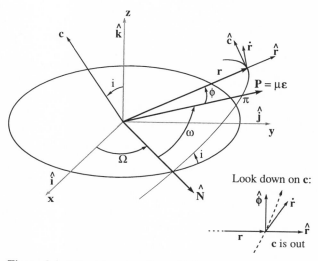

Figure 8.1 The geometry of an orbit in space.

8.5 Interpretation of $\hat{\mathbf{c}}$, $\hat{\boldsymbol{\varepsilon}}$, and $\hat{\mathbf{N}}$

From figures 8.1, 8.2, and Appendix A, we get the following equations for the unit vectors $\hat{\mathbf{c}}$, $\hat{\boldsymbol{\varepsilon}}$, and $\hat{\mathbf{N}}$:

$$\hat{\mathbf{c}} = \frac{1}{c}\mathbf{c} = \sin i \sin \Omega \,\hat{\mathbf{i}} - \sin i \cos \Omega \,\hat{\mathbf{j}} + \cos i \,\hat{\mathbf{k}}$$

$$\hat{\boldsymbol{\varepsilon}} = \frac{1}{\varepsilon}\boldsymbol{\varepsilon} = (\cos \Omega \cos \omega - \sin \Omega \sin \omega \cos i)\,\hat{\mathbf{i}}$$

$$+ (\sin \Omega \cos \omega + \cos \Omega \sin \omega \cos i)\,\hat{\mathbf{j}} + \sin \omega \sin i \,\hat{\mathbf{k}}$$

$$\hat{\mathbf{N}} \sin i = \hat{\mathbf{k}} \times \hat{\mathbf{c}}$$

$$\hat{\mathbf{N}} = \cos \Omega \,\hat{\mathbf{i}} + \sin \Omega \,\hat{\mathbf{j}}$$

8.6 The Perturbed Two-Body Problem

Perturbed two-body motion is described by the solution to the differential equation

$$\ddot{\mathbf{r}} + \frac{\mu}{r^3}\mathbf{r} = \mathbf{F}, \tag{8.34}$$

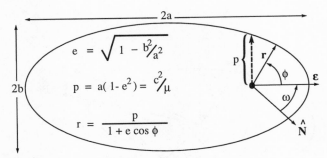

Figure 8.2 Geometry of the orbital plane.

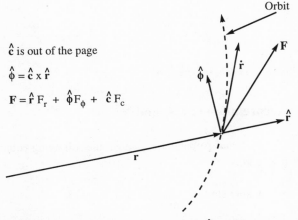

Figure 8.3 Perturbation $\mathbf{F}$ and the $\hat{\mathbf{c}}$, $\hat{\mathbf{r}}$, $\hat{\boldsymbol{\phi}}$ coordinate system.

where $\mathbf{F}$ is the perturbation. When $\mathbf{F} = 0$ we have the two-body problem as reviewed earlier. We now develop differential equations for some of the orbital elements.

8.6.1 Energy and Semi-Major Axis

The energy is

$$h = \frac{1}{2}\dot{\mathbf{r}} \cdot \dot{\mathbf{r}} - \frac{\mu}{r}. \tag{8.35}$$

When $\mathbf{F} = 0$, h is an integral (constant) of the motion (fig. 8.3). When $\mathbf{F} \neq 0$, h can become a variable depending on the nature of $\mathbf{F}$. Applying Poisson's method to the energy gives

$$\dot{h} = \frac{\partial h}{\partial \dot{\mathbf{r}}} \mathbf{F} \qquad (8.36)$$

$$= \dot{\mathbf{r}} \cdot \mathbf{F}. \qquad (8.37)$$

Note that here we have considered $\frac{\partial h}{\partial \dot{\mathbf{r}}}$ to be a row vector until the last step. Also note that $\frac{\partial}{\partial \dot{\mathbf{r}}}(\frac{\mu}{r}) = 0$. We can also obtain this well-known result by differentiating equation (8.35) with respect to time:

$$\dot{h} = \frac{1}{2}(2\dot{\mathbf{r}} \cdot \ddot{\mathbf{r}}) + \frac{\mu}{r^2}\dot{r}.$$

Using the elementary derivative

$$\dot{r} = \frac{\mathbf{r} \cdot \dot{\mathbf{r}}}{r},$$

we obtain

$$\dot{h} = \dot{\mathbf{r}} \cdot \ddot{\mathbf{r}} + \frac{\mu}{r^2}\frac{\mathbf{r} \cdot \dot{\mathbf{r}}}{r}$$

$$= \dot{\mathbf{r}} \cdot \left(\ddot{\mathbf{r}} + \frac{\mu}{r^3}\mathbf{r}\right);$$

and now, using equation (8.34),

$$\dot{h} = \dot{\mathbf{r}} \cdot \mathbf{F},$$

which is the same result as equation (8.37), which we derived using Poisson's method. Since the energy and semi-major axis have the relationship

$$a = \frac{-\mu}{2h}, \qquad (8.38)$$

we can relate their derivatives as

$$\dot{a} = \frac{\mu}{2h^2}\dot{h}.$$

By using equation (8.38) we get the result entirely in terms of a:

$$\dot{a} = \frac{2}{\mu}a^2\dot{\mathbf{r}} \cdot \mathbf{F}. \qquad (8.39)$$

Note from equations (8.37) and (8.39) that h and a are *not* affected by a component of $\mathbf{F}$ which is *normal* to the velocity $\dot{\mathbf{r}}$. That is, if $\dot{\mathbf{r}} \perp \mathbf{F}$ then $\dot{\mathbf{r}} \cdot \mathbf{F} = 0$, which results in $\dot{h} = 0$ and $\dot{a} = 0$.

8.6.2 Angular Momentum

The angular momentum is

$$\mathbf{c} = \mathbf{r} \times \dot{\mathbf{r}}, \tag{8.40}$$

which is a constant vector if $\mathbf{F} = 0$. For $\mathbf{F} \neq 0$ we apply Poisson's method to get the differential equations

$$\dot{\mathbf{c}} = \left[\frac{\partial \mathbf{c}}{\partial \dot{\mathbf{r}}} \right] \mathbf{F}$$

$$= \left[\frac{\partial}{\partial \dot{\mathbf{r}}} (\mathbf{r} \times \dot{\mathbf{r}}) \right] \mathbf{F}$$

$$= (\mathbf{r} \times) \mathbf{F}$$

$$\dot{\mathbf{c}} = \mathbf{r} \times \mathbf{F}. \tag{8.41}$$

Note that in this development we have used the fact that a cross product can be represented as a matrix, as we discussed in §1.2, that is,

$$R = \frac{\partial}{\partial \dot{\mathbf{r}}} (\mathbf{r} \times \dot{\mathbf{r}}) = \mathbf{r} \times,$$

which is a 3×3 matrix of the form

$$R = \begin{pmatrix} 0 & -x_3 & x_2 \\ x_3 & 0 & -x_1 \\ -x_2 & x_1 & 0 \end{pmatrix},$$

where the cartesian coordinates of $\mathbf{r}$ in an inertial system are x_1, x_2, x_3. Note also from this result that $\mathbf{c}$ is *not* effected by a component of $\mathbf{F}$ that is parallel to $\mathbf{r}$.

Equation (8.41) can also be derived by differentiating equation (8.40) and using equation (8.34):

$$\dot{\mathbf{c}} = \dot{\mathbf{r}} \times \dot{\mathbf{r}} + \mathbf{r} \times \ddot{\mathbf{r}}$$

$$= \mathbf{r} \times \left(\mathbf{F} - \frac{\mu}{r^3} \mathbf{r} \right)$$

$$\dot{\mathbf{c}} = \mathbf{r} \times \mathbf{F}. \tag{8.42}$$

A differential equation for the *magnitude* of the angular momentum (*note:*

$\dot{c} \neq \|\dot{\mathbf{c}}\|$) can be obtained as follows:

$$c^2 = \mathbf{c} \cdot \mathbf{c}$$
$$2c\dot{c} = 2(\mathbf{c} \cdot \dot{\mathbf{c}})$$
$$c\dot{c} = \mathbf{c} \cdot (\mathbf{r} \times \mathbf{F})$$
$$\dot{c} = \frac{1}{c}\mathbf{c} \cdot \mathbf{r} \times \mathbf{F}. \qquad (8.43)$$

By expressing the perturbation in the $\hat{\mathbf{c}}, \hat{\mathbf{r}}, \hat{\boldsymbol{\phi}}$ coordinate system, we obtain

$$\dot{c} = r\hat{\boldsymbol{\phi}} \cdot \mathbf{F}, \qquad (8.44)$$

where we have used the relation $\hat{\mathbf{c}} \times \hat{\mathbf{r}} = \hat{\boldsymbol{\phi}}$ defined earlier. Note that c is affected only by the *horizontal* component of $\mathbf{F}$.

8.6.3 Inclination

The differential equation for the inclination follows from the equation for $\mathbf{c}$,

$$\mathbf{c} = c(\sin i \sin \Omega\, \hat{\mathbf{i}} - \sin i \cos \Omega\, \hat{\mathbf{j}} + \cos i\, \hat{\mathbf{k}}). \qquad (8.45)$$

Note that $c \cos i = \hat{\mathbf{k}} \cdot \mathbf{c}$, which we differentiate to obtain

$$\dot{c} \cos i - c \sin i \frac{di}{dt} = \hat{\mathbf{k}} \cdot \dot{\mathbf{c}}.$$

Solve for $\frac{di}{dt}$ and use equations (8.41) and (8.44) for $\dot{\mathbf{c}}$ and $\dot{c}$,

$$c \sin i \frac{di}{dt} = \dot{c} \cos i - \hat{\mathbf{k}} \cdot \dot{\mathbf{c}}$$
$$= r\hat{\boldsymbol{\phi}} \cdot \mathbf{F} \cos i - \hat{\mathbf{k}} \cdot \mathbf{r} \times \mathbf{F}.$$

Once again, using $\hat{\boldsymbol{\phi}} = \hat{\mathbf{c}} \times \hat{\mathbf{r}}$ we obtain

$$c \sin i \frac{di}{dt} = r\left[\hat{\mathbf{c}} \cos i - \hat{\mathbf{k}}\right] \cdot \hat{\mathbf{r}} \times \mathbf{F}.$$

Substituting for $\hat{\mathbf{c}}$ from Appendix A (also §8.5), we get

$$c \sin i \frac{di}{dt} = r\left[(\sin i \sin \Omega\, \hat{\mathbf{i}} - \sin i \cos \Omega\, \hat{\mathbf{j}} + \cos i\, \hat{\mathbf{k}}) \cos i - \hat{\mathbf{k}}\right] \cdot \hat{\mathbf{r}} \times \mathbf{F}.$$

Note the coefficient of $\hat{\mathbf{k}}$ causes a convenient cancelation of $\sin i$, which results in

$$c\frac{di}{dt} = r\left[\cos i \sin \Omega \,\hat{\mathbf{i}} - \cos i \cos \Omega \,\hat{\mathbf{j}} - \sin i \,\hat{\mathbf{k}}\right] \cdot \hat{\mathbf{r}} \times \mathbf{F}. \qquad (8.46)$$

The term in brackets can now be shown to be $\hat{\mathbf{N}} \times \hat{\mathbf{c}}$. From Appendix A we get $\hat{\mathbf{c}}$ and

$$\hat{\mathbf{N}} = \cos \Omega \,\hat{\mathbf{i}} + \sin \Omega \,\hat{\mathbf{j}}. \qquad (8.47)$$

Therefore, using the "determinant" method for evaluating a cross product,

$$\hat{\mathbf{N}} \times \hat{\mathbf{c}} = \begin{vmatrix} \hat{\mathbf{i}} & \hat{\mathbf{j}} & \hat{\mathbf{k}} \\ \cos \Omega & \sin \Omega & 0 \\ \sin i \sin \Omega & -\sin i \cos \Omega & \cos i \end{vmatrix}$$

we arrive at

$$\hat{\mathbf{N}} \times \hat{\mathbf{c}} = \sin \Omega \cos i \,\hat{\mathbf{i}} - \cos \Omega \cos i \,\hat{\mathbf{j}} - \sin i \,\hat{\mathbf{k}}, \qquad (8.48)$$

which is the same as the bracketed term in equation (8.46). Substituting equation (8.48) into equation (8.46) results in

$$c\frac{di}{dt} = r\,(\hat{\mathbf{N}} \times \hat{\mathbf{c}}) \cdot (\hat{\mathbf{r}} \times \mathbf{F}).$$

Expand the right side using the elementary vector identity already given in §2.4.4 of chapter 2 to arrive at

$$c\frac{di}{dt} = r\left[(\hat{\mathbf{c}} \cdot \mathbf{F})(\hat{\mathbf{N}} \cdot \hat{\mathbf{r}}) - (\hat{\mathbf{c}} \cdot \hat{\mathbf{r}})(\hat{\mathbf{N}} \cdot \mathbf{F})\right].$$

Using $\hat{\mathbf{c}} \cdot \hat{\mathbf{r}} = 0$ and $\hat{\mathbf{r}} \cdot \hat{\mathbf{N}} = \cos(\omega + \phi)$ (see fig. 8.1), we obtain finally

$$\frac{di}{dt} = \frac{r}{c}\cos(\omega + \phi)\hat{\mathbf{c}} \cdot \mathbf{F}. \qquad (8.49)$$

Only a perturbation *normal* to the orbital plane affects the inclination.

8.6.4 The Node Angle Ω

The equation for the ascending node angle Ω is (refer to §4.6)

$$\tan \Omega = \frac{\hat{\mathbf{i}} \cdot \mathbf{c}}{-\hat{\mathbf{j}} \cdot \mathbf{c}},$$

where, from equation (8.45), we have

$$\hat{\mathbf{i}} \cdot \mathbf{c} = c \sin i \sin \Omega$$

$$\hat{\mathbf{j}} \cdot \mathbf{c} = -c \sin i \cos \Omega.$$

Taking the derivative of $\tan \Omega$ with respect to time,

$$-\sec^2 \Omega \, \dot{\Omega} = \frac{(\hat{\mathbf{j}} \cdot \mathbf{c})(\hat{\mathbf{i}} \cdot \dot{\mathbf{c}}) - (\hat{\mathbf{i}} \cdot \mathbf{c})(\hat{\mathbf{j}} \cdot \dot{\mathbf{c}})}{(\hat{\mathbf{j}} \cdot \mathbf{c})^2}.$$

Now multiply by $(\hat{\mathbf{j}} \cdot \mathbf{c})^2$ and use equation (8.45) to get

$$\frac{\dot{\Omega}}{\cos^2 \Omega} \left(c^2 \sin^2 i \, \cos^2 \Omega \right) = \left(c \sin i \cos \Omega \, \hat{\mathbf{i}} + c \sin i \sin \Omega \, \hat{\mathbf{j}} \right) \cdot \dot{\mathbf{c}}.$$

Noting the cancelations we get

$$\dot{\Omega} \, c \sin i = \left(\cos \Omega \, \hat{\mathbf{i}} + \sin \Omega \, \hat{\mathbf{j}} \right) \cdot \dot{\mathbf{c}}.$$

Using equation (8.41) for $\dot{\mathbf{c}}$, equation (A.2) for $\hat{\mathbf{N}}$, and using $\hat{\mathbf{N}} \times \hat{\mathbf{r}} = \hat{\mathbf{c}} \sin(\omega + \phi)$ from figure 8.1,

$$\dot{\Omega} \, c \sin i = \hat{\mathbf{N}} \cdot \mathbf{r} \times \mathbf{F}$$

$$= r \, \hat{\mathbf{N}} \times \hat{\mathbf{r}} \cdot \mathbf{F}. \tag{8.50}$$

Finally, we obtain

$$\dot{\Omega} = \frac{r}{c \sin i} \sin(\omega + \phi) \hat{\mathbf{c}} \cdot \mathbf{F}. \tag{8.51}$$

Only perturbations normal to the orbital plane affect the ascending node angle Ω.

8.6.5 Laplace Vector

Here we introduce the change in notation noted earlier,

$$\mathbf{P} = \mu \boldsymbol{\varepsilon}.$$

From §8.4 the Laplace vector in terms of position and velocity is

$$\mu \boldsymbol{\varepsilon} = -\frac{\mu}{r} \mathbf{r} - \mathbf{c} \times \dot{\mathbf{r}}, \tag{8.52}$$

which is a *constant* vector when $\mathbf{F} = 0$. For the case where $\mathbf{F} \neq 0$ we use Poisson's method to get a differential equation,

$$\mu \dot{\varepsilon} = \left[\frac{\partial (\mu \varepsilon)}{\partial \dot{\mathbf{r}}} \right] \mathbf{F}$$

$$= -\left[\frac{\partial}{\partial \dot{\mathbf{r}}} (\mathbf{c} \times \dot{\mathbf{r}}) \right] \mathbf{F}.$$

Note that here we are considering $\frac{\partial}{\partial \dot{\mathbf{r}}} (\mathbf{c} \times \dot{\mathbf{r}})$ as a matrix,

$$\mu \dot{\varepsilon} = \left[\dot{\mathbf{r}} \times \frac{\partial \mathbf{c}}{\partial \dot{\mathbf{r}}} - \mathbf{c} \times \right] \mathbf{F}.$$

From the development of $\dot{\mathbf{c}}$, equation (8.41), we have

$$\dot{\mathbf{c}} = \left[\frac{\partial \mathbf{c}}{\partial \dot{\mathbf{r}}} \right] \mathbf{F} = \mathbf{r} \times \mathbf{F},$$

and so

$$\mu \dot{\varepsilon} = \dot{\mathbf{r}} \times (\mathbf{r} \times \mathbf{F}) - (\mathbf{r} \times \dot{\mathbf{r}}) \times \mathbf{F}. \tag{8.53}$$

Expanding the two triple vector products we obtain

$$\mu \dot{\varepsilon} = (\dot{\mathbf{r}} \cdot \mathbf{F})\mathbf{r} - (\mathbf{r} \cdot \dot{\mathbf{r}})\mathbf{F} - [(\mathbf{r} \cdot \mathbf{F})\dot{\mathbf{r}} - (\dot{\mathbf{r}} \cdot \mathbf{F})\mathbf{r}],$$

which simplifies to

$$\mu \dot{\varepsilon} = 2(\dot{\mathbf{r}} \cdot \mathbf{F})\mathbf{r} - (\mathbf{r} \cdot \dot{\mathbf{r}})\mathbf{F} - (\mathbf{r} \cdot \mathbf{F})\dot{\mathbf{r}}. \tag{8.54}$$

8.7 Some Partially Solved Problems

In some important cases a perturbation will have some of the integrals or constants unaffected. In other cases new (non-two-body) integrals appear. In the following sections we consider a few examples.

8.7.1 Conservative Potentials

Assume a potential of the form $V(\mathbf{r})$, that is, a potential that depends only on position. The differential equation of motion equation (8.34) becomes

$$\ddot{\mathbf{r}} + \frac{\mu}{r^3}\mathbf{r} = \mathbf{F} = -\frac{\partial V}{\partial \mathbf{r}}.$$

From equation (8.37) the differential equation for the energy is

$$\dot{h} = \dot{\mathbf{r}} \cdot \mathbf{F} = -\dot{\mathbf{r}} \cdot \frac{\partial V}{\partial \mathbf{r}}.$$

The total derivative of V is

$$\frac{d}{dt}(V) = \frac{\partial V}{\partial \mathbf{r}} \cdot \dot{\mathbf{r}} + \frac{\partial V}{\partial t}.$$

Note that $V(\mathbf{r})$ is dependent on $\mathbf{r}(t)$ but t does not appear *explicitly* in V. Therefore for this situation,

$$\frac{\partial V}{\partial t} = 0.$$

The differential equation for energy becomes

$$\dot{h} = -\frac{d}{dt}(V),$$

which we integrate to get

$$h + V(\mathbf{r}) = constant = \text{total energy}.$$

For the case where the potential is a function of position only, the total energy is conserved (Stiefel and Scheifele [64]). The potential $V(\mathbf{r})$ is called a *conservative potential*.

8.7.2 Oblate Planet Potential

Planets typically have excessive mass in a "belt" around their equator. This mass is distributed symmetrically about the direction ($\hat{\mathbf{k}}$) normal to the planet's equator. The problem is to find the integrals of motion.

The equation of motion (as before) is

$$\ddot{\mathbf{r}} + \frac{\mu}{r^3}\mathbf{r} = \mathbf{F} = -\frac{\partial V}{\partial \mathbf{r}},$$

where the potential that is a function of position only is given by

$$V = V(\mathbf{r}) = \frac{\epsilon}{r^3}\left(3 \sin^2 \delta - 1\right).$$

We call ϵ the perturbing parameter where

$$\epsilon = \frac{1}{2}\mu J_2 a_e^2.$$

The a_e term is the equatorial radius of the planet and the J_2 term is the oblateness coefficient, which is on the order of 10^{-3} for the Earth. The angle δ is the declination (latitude) of $\mathbf{r}$, and $\sin \delta = \hat{\mathbf{k}} \cdot \hat{\mathbf{r}}$. The gradient of the potential is

$$\frac{\partial V}{\partial \mathbf{r}} = \frac{\partial V}{\partial r} \frac{\partial r}{\partial \mathbf{r}} + \frac{\partial V}{\partial (\sin \delta)} \frac{\partial (\sin \delta)}{\partial \mathbf{r}}.$$

For $\frac{\partial r}{\partial \mathbf{r}}$ the partials are found as follows. Since

$$r^2 = \mathbf{r} \cdot \mathbf{r},$$

we have

$$\frac{\partial r}{\partial \mathbf{r}} = \frac{1}{r} \mathbf{r} = \hat{\mathbf{r}},$$

and for $\frac{\partial (\sin \delta)}{\partial \mathbf{r}}$, since

$$r \sin \delta = \hat{\mathbf{k}} \cdot \mathbf{r},$$

the partial derivative yields

$$\frac{\partial r}{\partial \mathbf{r}} \sin \delta + r \frac{\partial (\sin \delta)}{\partial \mathbf{r}} = \hat{\mathbf{k}} \frac{\partial r}{\partial \mathbf{r}}.$$

Using the above result for $\frac{\partial r}{\partial \mathbf{r}}$ we get

$$\frac{\partial (\sin \delta)}{\partial \mathbf{r}} = \frac{1}{r} \left(\hat{\mathbf{k}} - \hat{\mathbf{r}} \sin \delta \right).$$

So the gradient of the potential becomes

$$\frac{\partial V}{\partial \mathbf{r}} = \hat{\mathbf{r}} \left(\frac{\partial V}{\partial r} - \frac{1}{r} \frac{\partial V}{\partial (\sin \delta)} \sin \delta \right) + \hat{\mathbf{k}} \frac{1}{r} \frac{\partial V}{\partial (\sin \delta)},$$

or by making some obvious abbreviations for the coefficients of $\hat{\mathbf{r}}$ and $\hat{\mathbf{k}}$,

$$\frac{\partial V}{\partial \mathbf{r}} = f_1 \hat{\mathbf{r}} + f_3 \hat{\mathbf{k}}.$$

We can now derive an integral of the motion. First observe that

$$\mathbf{F} = -\frac{\partial V}{\partial \mathbf{r}} = -f_1 \hat{\mathbf{r}} - f_3 \hat{\mathbf{k}}.$$

Now recall that the differential equation for the angular momentum (eq. 8.41) is

$$\dot{\mathbf{c}} = \mathbf{r} \times \mathbf{F}.$$

For this case the first term vanishes and we dot with $\hat{\mathbf{k}}$ to obtain

$$\hat{\mathbf{k}} \cdot \dot{\mathbf{c}} = -\hat{\mathbf{k}} \cdot \mathbf{r} \times \hat{\mathbf{k}} \, f_3 = 0.$$

Since $\hat{\mathbf{k}}$ is constant,

$$\frac{d}{dt}(\hat{\mathbf{k}} \cdot \mathbf{c}) = 0.$$

So from equation (8.45),

$$\hat{\mathbf{k}} \cdot \mathbf{c} = c \cos i = constant.$$

This result shows that the component of the angular momentum along the symmetry axis ($\hat{\mathbf{k}}$) is *constant*. Also, since $V = V(\mathbf{r})$, we note from the previous section that the total energy is conserved,

$$h + V = constant.$$

The oblateness problem has two integrals of the motion. We note here that this result can be generalized for any planetary potential that has only zonal terms (see §11.2), that is, terms that have no longitudinal dependence.

8.7.3 Time-Dependent Potential

For this case we observe that in addition to the *implicit* dependence on time through the position $\mathbf{r}(t)$, the potential has an *explicit* dependence on time.

The differential equation of motion (8.34) becomes

$$\ddot{\mathbf{r}} + \frac{\mu}{r^3}\mathbf{r} = \mathbf{F} = -\frac{\partial V(\mathbf{r}, t)}{\partial \mathbf{r}}. \tag{8.55}$$

In the following discussion we will simply write V, and this is understood to be $V(\mathbf{r}, t)$. For this potential, can we find an integral of the motion? Bond and Gottlieb [18] submit the following approach to answer this question.

The differential equation for the energy (8.37) becomes

$$\dot{h} = \dot{\mathbf{r}} \cdot \mathbf{F} = -\dot{\mathbf{r}} \cdot \frac{\partial V}{\partial \mathbf{r}}.$$

The total derivative of V is

$$\frac{d}{dt}(V) = \frac{\partial V}{\partial \mathbf{r}} \cdot \dot{\mathbf{r}} + \frac{\partial V}{\partial t}.$$

Using this in the differential equation for the energy, we eliminate $\dot{\mathbf{r}} \cdot \frac{\partial V}{\partial \mathbf{r}}$ to obtain

$$\dot{h} = -\frac{d}{dt}(V) + \frac{\partial V}{\partial t}$$

or

$$\frac{d}{dt}(h + V) = \frac{\partial V}{\partial t}. \tag{8.56}$$

The left side of this equation can be integrated directly, but the right side requires special attention. For certain cases, as shown in the following two subsections,

$$\frac{\partial V}{\partial t} = -\boldsymbol{\omega} \cdot \mathbf{r} \times \frac{\partial V}{\partial \mathbf{r}}, \tag{8.57}$$

where $\boldsymbol{\omega}$ is a constant vector. Equation (8.57) can be made an integrable function by crossing equation (8.55) with the position vector $\mathbf{r}$ and dotting the result with $\boldsymbol{\omega}$. The first operation gives

$$\mathbf{r} \times \left(\ddot{\mathbf{r}} + \frac{\mu}{r^3}\mathbf{r} \right) = -\mathbf{r} \times \frac{\partial V}{\partial \mathbf{r}},$$

which can be reduced to

$$\frac{d}{dt}(\mathbf{r} \times \dot{\mathbf{r}}) = -\mathbf{r} \times \frac{\partial V}{\partial \mathbf{r}}.$$

Now take the dot product of this equation with $\boldsymbol{\omega}$ to get

$$\boldsymbol{\omega} \cdot \frac{d}{dt}(\mathbf{r} \times \dot{\mathbf{r}}) = -\boldsymbol{\omega} \cdot \mathbf{r} \times \frac{\partial V}{\partial \mathbf{r}}. \tag{8.58}$$

Substituting equation (8.57) in the right side of equation (8.58) and exchanging sides results in

$$\frac{\partial V}{\partial t} = \boldsymbol{\omega} \cdot \frac{d}{dt}(\mathbf{r} \times \dot{\mathbf{r}}). \tag{8.59}$$

Now substitute (8.59) into equation (8.56) to get

$$\frac{d}{dt}(h + V) = \boldsymbol{\omega} \cdot \frac{d}{dt}(\mathbf{r} \times \dot{\mathbf{r}}),$$

which can now be integrated to give

$$h + V - \boldsymbol{\omega} \cdot \mathbf{r} \times \dot{\mathbf{r}} = constant. \tag{8.60}$$

Recognizing that $\mathbf{c} = \mathbf{r} \times \dot{\mathbf{r}} = $ the angular momentum, equation (8.60) becomes

$$h + V(\mathbf{r}, t) - \boldsymbol{\omega} \cdot \mathbf{c} = constant. \tag{8.61}$$

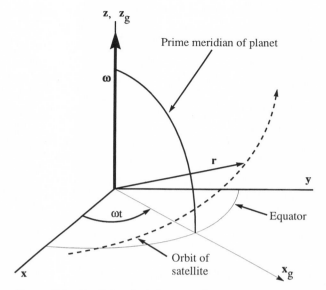

Figure 8.4 Inertial and rotating coordinate systems.

Note that for this type of potential neither the total energy nor the angular momentum is *constant*. However, the total energy minus the component of angular momentum along $\boldsymbol{\omega}$ *is constant*. We shall call this constant the *Jacobi integral* because of its analogy to the Jacobi integral of the Restricted Three-Body Problem (Szebehely [69]). The integral in the form of equation (8.61) was used by Nacozy [57], who referred to it as being analogous to the Jacobi integral of the Restricted Three-Body Problem.

8.7.4 Derivatives for a Rotating Planet

As in the report of Bond and Mulcihy [22], assume a nonsymmetrical planet that is rotating at constant angular velocity $\boldsymbol{\omega}$. As the planet rotates, the potential changes explicitly with time. So the potential experienced by a satellite is a function of the angular position of the planet as well as the position of the satellite. Let x, y, z be the *inertial* system and let x_g, y_g, z_g be the *rotating* system *fixed* in the planet (fig. 8.4). The z and z_g axes are collinear (along $\boldsymbol{\omega}$). Now the position vector is the same in either system,

$$\mathbf{r} = \mathbf{r}_g, \tag{8.62}$$

where $\mathbf{r}$ is in the *inertial* system (x, y, z) and $\mathbf{r}_g$ is in the *rotating* system (x_g, y_g, z_g). Although the vectors $\mathbf{r}$ and $\mathbf{r}_g$ are the same, the components of the vectors are related by the vector-matrix relation

$$\mathbf{r}_g = \begin{pmatrix} \cos\omega t & \sin\omega t & 0 \\ -\sin\omega t & \cos\omega t & 0 \\ 0 & 0 & 1 \end{pmatrix} \mathbf{r},$$

which is derived in Appendix A. The velocity vectors are related by

$$\dot{\mathbf{r}} = \dot{\mathbf{r}}_g + \boldsymbol{\omega} \times \mathbf{r}. \tag{8.63}$$

The potential function (a scalar) has the same numerical value in both systems but has different forms:

$$V(\mathbf{r}, t) = V(\mathbf{r}_g). \tag{8.64}$$

Now take the total derivative of the potential in each system. In the inertial system,

$$\frac{d}{dt} V(\mathbf{r}, t) = \frac{\partial V(\mathbf{r}, t)}{\partial \mathbf{r}} \cdot \dot{\mathbf{r}} + \frac{\partial V(\mathbf{r}, t)}{\partial t}, \tag{8.65}$$

and in the rotating system,

$$\frac{d}{dt} V(\mathbf{r}_g) = \frac{\partial V(\mathbf{r}_g)}{\partial \mathbf{r}_g} \cdot \dot{\mathbf{r}}_g. \tag{8.66}$$

The values of the two total derivatives are the same. We therefore equate equations (8.65) and (8.66), at the same time using equation (8.63):

$$\frac{\partial V(\mathbf{r}, t)}{\partial \mathbf{r}} \cdot (\dot{\mathbf{r}}_g + \boldsymbol{\omega} \times \mathbf{r}) + \frac{\partial V(\mathbf{r}, t)}{\partial t} = \frac{\partial V(\mathbf{r}_g)}{\partial \mathbf{r}_g} \cdot \dot{\mathbf{r}}_g.$$

Since $\mathbf{r} = \mathbf{r}_g$, the partial with respect to position terms cancels. We have, therefore,

$$\frac{\partial V(\mathbf{r}, t)}{\partial t} = -\boldsymbol{\omega} \times \mathbf{r} \cdot \frac{\partial V(\mathbf{r}, t)}{\partial \mathbf{r}}.$$

Interchange the cross and dot products to get

$$\frac{\partial V(\mathbf{r}, t)}{\partial t} = -\boldsymbol{\omega} \cdot \mathbf{r} \times \frac{\partial V(\mathbf{r}, t)}{\partial \mathbf{r}},$$

which has the form of equation (8.57).

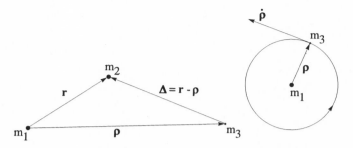

Figure 8.5   The two-body system $(m_1 + m_2)$ and the perturbing third mass (m_3).

8.7.5 Derivatives for Perturbation by a Third Body

As in the report by Bond and Gottlieb [18], assume a third body of mass m_3 that is orbiting one of the masses (m_1) of a perturbed two-body system (fig. 8.5). The differential equation of this perturbed system is

$$\ddot{\mathbf{r}} + \frac{\mu}{r^3}\mathbf{r} = -\frac{\partial V(\mathbf{r}, t)}{\partial \mathbf{r}}, \tag{8.67}$$

where $\mu = G(m_1 + m_2)$ and m_3 is the mass of the satellite.

The third body potential function is

$$V = V(\mathbf{r}, t) = -Gm_3 \left[\frac{1}{\Delta} - \frac{\mathbf{r} \cdot \boldsymbol{\rho}}{\rho^3} \right], \tag{8.68}$$

with the gradient

$$\frac{\partial V}{\partial \mathbf{r}} = Gm_3 \left[\frac{\mathbf{r} - \boldsymbol{\rho}}{\Delta^3} + \frac{\boldsymbol{\rho}}{\rho^3} \right], \tag{8.69}$$

as shown in chapter 11, equation (11.13). Note that the explicit time dependence of the potential occurs through the position of the planet. That is,

$$\boldsymbol{\rho} = \boldsymbol{\rho}(t). \tag{8.70}$$

Now we assume that m_3 is in a circular orbit,

$$\rho = \|\boldsymbol{\rho}\| = constant,$$

and therefore

$$\boldsymbol{\rho} \cdot \dot{\boldsymbol{\rho}} = 0.$$

Since $\rho = \rho(t)$, we have that

$$\frac{d}{dt}\rho = \frac{\partial\rho(t)}{\partial t} = \dot{\rho}. \tag{8.71}$$

Thus from equation (8.68),

$$\frac{\partial V}{\partial t} = -Gm_3\left[\frac{-1}{\Delta^2}\frac{\partial\Delta}{\partial t} - \frac{1}{\rho^3}\mathbf{r}\cdot\frac{\partial\rho}{\partial t} + \frac{3\mathbf{r}\cdot\rho}{\rho^4}\frac{\partial\rho}{\partial t}\right]. \tag{8.72}$$

But

$$\frac{\partial\rho}{\partial t} = 0,$$

since $\rho = constant$. Also,

$$\Delta^2 = (\mathbf{r} - \rho)\cdot(\mathbf{r} - \rho).$$

So the partial derivative of Δ becomes

$$2\Delta\frac{\partial\Delta}{\partial t} = -2(\mathbf{r} - \rho)\cdot\frac{\partial\rho}{\partial t} = -2\mathbf{r}\cdot\frac{\partial\rho}{\partial t}.$$

Substitute this back into equation (8.72):

$$\frac{\partial V}{\partial t} = -Gm_3\left[\frac{1}{\Delta^3}\mathbf{r}\cdot\frac{\partial\rho}{\partial t} - \frac{1}{\rho^3}\mathbf{r}\cdot\frac{\partial\rho}{\partial t}\right]. \tag{8.73}$$

Using equation (8.71) this becomes

$$\frac{\partial V}{\partial t} = -Gm_3(\mathbf{r}\cdot\dot{\rho})\left[\frac{1}{\Delta^3} - \frac{1}{\rho^3}\right]. \tag{8.74}$$

Now we define the constant vector $\boldsymbol{\omega}$,

$$\boldsymbol{\omega} = n_3\hat{\mathbf{c}}_3, \tag{8.75}$$

where n_3 is the mean motion of m_3,

$$n_3^2 = \frac{G(m_1 + m_3)}{\rho^3} = constant. \tag{8.76}$$

Also, $\hat{\mathbf{c}}_3$ is normal to the plane of motion of the planet m_3, which is the same as the unit angular momentum of m_3,

$$\hat{\mathbf{c}}_3 = \frac{\rho\times\dot{\rho}}{\|\rho\times\dot{\rho}\|}. \tag{8.77}$$

Now because of the circular orbit assumption for m_3,

$$\dot{\rho} = \boldsymbol{\omega}\times\rho. \tag{8.78}$$

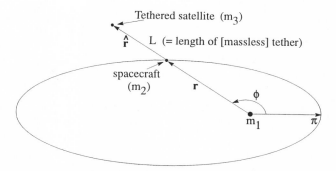

Figure 8.6 Tethered satellite system.

So the factor $\mathbf{r} \cdot \dot{\rho}$ in equation (8.74) becomes

$$\mathbf{r} \cdot \dot{\rho} = \mathbf{r} \cdot \boldsymbol{\omega} \times \rho = -\boldsymbol{\omega} \cdot \mathbf{r} \times \rho, \tag{8.79}$$

and equation (8.74) then is

$$\frac{\partial V}{\partial t} = G m_3 \left(\boldsymbol{\omega} \cdot \mathbf{r} \times \rho \right) \left[\frac{1}{\Delta^3} - \frac{1}{\rho^3} \right], \tag{8.80}$$

but from the gradient equation (8.69),

$$\mathbf{r} \times \frac{\partial V}{\partial \mathbf{r}} = G m_3 \left[\frac{-\mathbf{r} \times \rho}{\Delta^3} + \frac{\mathbf{r} \times \rho}{\rho^3} \right]$$

and

$$\boldsymbol{\omega} \cdot \mathbf{r} \times \frac{\partial V}{\partial \mathbf{r}} = G m_3 \boldsymbol{\omega} \cdot \mathbf{r} \times \rho \left[\frac{-1}{\Delta^3} + \frac{1}{\rho^3} \right]. \tag{8.81}$$

By comparison of equations (8.80) and (8.81),

$$\frac{\partial V}{\partial t} = -\boldsymbol{\omega} \cdot \mathbf{r} \times \frac{\partial V}{\partial \mathbf{r}}, \tag{8.82}$$

which has the form of equation (8.57).

8.7.6 Tethered Satellite Problem

For this problem (Bond [16]) we make the assumptions that the tether is mass-less, the tether is deployed along the local vertical ($\hat{\mathbf{r}}$), and the orbit of the spacecraft is elliptical (fig. 8.6). The equations of motion are

$$\ddot{\mathbf{r}} + \frac{\mu}{r^3} \mathbf{r} = \mathbf{F} = \pm \frac{F_l}{m_2} \hat{\mathbf{r}},$$

where the positive force is directed away from the planet (m_1). Also,

$$\frac{F_I}{m_2} = \frac{\tilde{m}}{m_2}\frac{\mu L}{r^3}\left(\frac{c^2}{\mu r} + 2\right) = f(r),$$

where m_2 is the mass of the spacecraft, m_3 is the mass of the tethered satellite, and

$$\tilde{m} = \frac{m_3 m_2}{m_3 + m_2}$$

is the reduced mass of the combined spacecraft and tethered satellite system.

What are the integrals of the motion? Recall that for the angular momentum,

$$\dot{\mathbf{c}} = \mathbf{r} \times \mathbf{F}.$$

For this case, $\mathbf{F} = \pm f(r)\,\hat{\mathbf{r}}$, so

$$\dot{\mathbf{c}} = \mathbf{r} \times (\pm f(r)\,\hat{\mathbf{r}}) = 0.$$

Therefore, $\mathbf{c} = constant$. This implies that the plane is fixed in space, so

$$i = constant$$

$$\Omega = constant.$$

It also implies that the magnitude of the angular momentum (c) is constant. For the energy,

$$\dot{h} = \dot{\mathbf{r}} \cdot \mathbf{F},$$

which, for this case,

$$\dot{h} = \dot{\mathbf{r}} \cdot (\pm f(r)\,\hat{\mathbf{r}}) = \pm f(r)\frac{1}{r}\dot{\mathbf{r}} \cdot \mathbf{r}$$

$$\dot{h} = \pm f(r)\dot{r}.$$

Integrate to obtain (note that $f(r)$ is integrable)

$$h = \pm \int f(r)\,dr = \pm g(r) + constant.$$

Since h is the two-body energy, we can interpret the constant to be the total energy. We now investigate the behavior of a and e. Since $\mathbf{c} = constant$, the semi-latus rectum is also constant,

$$p = \frac{c^2}{\mu} \equiv a(1 - e^2) = constant.$$

Recall the equation relating the semi-major axis to the two-body energy,

$$a = -\frac{\mu}{2h}.$$

Since the two-body energy is not constant for this problem, a is *not* a constant. Also, since p is a constant, we see that e *is also not a constant*. Note, however, that they change in such a way that the semi-latus rectum *remains* constant. The tethered satellite problem that is deployed along the local vertical has four integrals of motion, which are the angular momentum vector and the total energy. This result is valid for any purely radial perturbation.

8.7.7 Drag Problem

This example is taken from Pollard [58]. Consider the differential equation of motion

$$\ddot{\mathbf{r}} + \frac{\mu}{r^3}\mathbf{r} = \mathbf{F} = -q\,\dot{\mathbf{r}}, \qquad q > 0.$$

The drag ($\mathbf{F} = -q\,\dot{\mathbf{r}}$) is a dissipative acceleration opposite to the velocity and acts continuously on a spacecraft.

Again, what are the integrals of the motion? From Poisson's method,

$$\dot{h} = \dot{\mathbf{r}} \cdot \mathbf{F} = -q\,\dot{\mathbf{r}} \cdot \dot{\mathbf{r}}.$$

Also, since

$$h = -\frac{\mu}{2a},$$

we have

$$\dot{h} = \frac{\mu}{2a^2}\dot{a}.$$

Solving for $\dot{a}$ we obtain

$$\dot{a} = -\frac{2a^2}{\mu}q\,\dot{\mathbf{r}} \cdot \dot{\mathbf{r}}.$$

The equations for $\dot{h}$ and $\dot{a}$ indicate that the energy and semi-major axis must always *decrease*.

The differential equation for the angular momentum is

$$\dot{\mathbf{c}} = \mathbf{r} \times \mathbf{F},$$

which for this case becomes

$$\dot{\mathbf{c}} = \mathbf{r} \times (-q\,\dot{\mathbf{r}})$$

$$\dot{\mathbf{c}} = -q\,\mathbf{c}.$$

For the magnitude of $\mathbf{c}$, differentiate the equation $c^2 = \mathbf{c} \cdot \mathbf{c}$ to get

$$2c\dot{c} = 2\mathbf{c} \cdot \dot{\mathbf{c}}.$$

Using the above result for $\dot{\mathbf{c}}$,

$$c\dot{c} = -q\,\mathbf{c} \cdot \mathbf{c} = -qc^2$$

$$\dot{c} = -qc.$$

We will now show that the plane of motion remains fixed. Since

$$\mathbf{c} = c\,\hat{\mathbf{c}}$$

$$\dot{\mathbf{c}} = \dot{c}\,\hat{\mathbf{c}} + c\,\frac{d\hat{\mathbf{c}}}{dt}.$$

Now solve for $\frac{d\hat{\mathbf{c}}}{dt}$ and use $\dot{\mathbf{c}}$ and $\dot{c}$ from above:

$$c\frac{d\hat{\mathbf{c}}}{dt} = \dot{\mathbf{c}} - \dot{c}\,\hat{\mathbf{c}} = -q\,\mathbf{c} + qc\,\hat{\mathbf{c}} = 0.$$

Therefore,

$$\frac{d\hat{\mathbf{c}}}{dt} = 0,$$

which upon integration becomes

$$\hat{\mathbf{c}} = constant.$$

Recall that $\hat{\mathbf{c}}$ is normal to the plane of motion. Since this normal vector is constant, the plane of motion remains fixed; therefore both the inclination i and the ascending node angle Ω are constant.

9

SPECIAL PERTURBATION METHODS

In this chapter we introduce the Sperling-Burdet approach to the perturbed two-body problem.

9.1 Propagation Error

A special perturbation method is a numerical approach for solving the perturbed two-body problem. In chapter 8 we briefly referred to two such methods, those of Cowell and Encke. Neither of these methods uses the orbital elements as the new dependent variables but instead they integrate the differential equations of motion in the cartesian coordinates or, in the case of Encke's method, in the deviation of the coordinates from the two-body solution. The propagation error (PE) associated with Cowell's method is the difference between two neighboring solutions of

$$\ddot{\mathbf{r}} + \frac{\mu}{r^3}\mathbf{r} = \mathbf{F},$$

having slightly different initial conditions. The difference between these two solutions can be found from the variational equation [15],

$$\frac{d}{dt}(\mathbf{PE}) = A\,\mathbf{PE}.$$

The perturbation error is a vector of the variations on position and velocity,

$$\mathbf{PE}^T = (\delta\mathbf{r}^T, \delta\dot{\mathbf{r}}^T).$$

By setting the perturbation $\mathbf{F} = 0$, the matrix A is

$$A = \begin{bmatrix} O & I \\ C & O \end{bmatrix},$$

where O is the 3×3 null matrix, I is the 3×3 identity matrix, and

$$C = -\frac{\mu}{r^3}\left[I - \mathbf{r}\,\mathbf{r}^T\right].$$

The solution of the variational equation is

$$\mathbf{PE}(\Delta t) = \left[e^{\int_0^{\Delta t} A dt}\right]\mathbf{PE}(0).$$

For a small-step Δt, we can consider the matrix A to be constant, and the solution becomes

$$\mathbf{PE}(\Delta t) = e^{A \Delta t}\mathbf{PE}(0).$$

Thus the propagation error over the step Δt is determined by the eigenvalues of the matrix A. Since one of the eigenvalues of A [15] is positive and real, having the value $\sqrt{2\mu/r^3}$, the propagation error will always increase. That is,

$$\|\mathbf{PE}(\Delta t)\| > \|\mathbf{PE}(0)\|.$$

The Encke method [15] exhibits the same characteristics; the propagation error will always increase.

Another very important point here is that the increase in propagation error for the Cowell and Encke methods occurs even when there are no perturbations. In this chapter we will develop a set of differential equations for the orbital elements that have no increase in propagation error when $\mathbf{F} = 0$.

9.2 Regularization

In 1961 Sperling [61] published his paper "Computation of Keplerian Conic Sections." In this paper he succeeded in the regularization and the linearization of the two-body problem as defined by equation (2.8). He accomplished this feat by the following steps: changing the independent variable from time (t) to fictitious time (s) by Sundmann's [67] transformation, and then embedding the Laplace vector and the Keplerian energy. This transformed the nonlinear two-body problem into a linear differential equation that could be readily solved as a harmonic oscillator. Following this development, Burdet [25] in 1968 published a perturbation theory based on Sperling's solution. Burdet, using the variation-of-parameters method, developed differential equations for the elements (or integrals of the motion), which appeared in Sperling's equations. This section will follow the Sperling-Burdet development beginning with the first step of regularization.

Regularization is defined as the elimination of singularities from the differential equations of motion. Note that the purpose is to regularize the differential equations and *not* their solution. We use an example from Stiefel and Scheifele [64] to illustrate:

$$x'' - \frac{\sin s}{1 - \cos s} x' + \frac{x}{1 - \cos s} = 0,$$

where $x' = dx/ds$. This equation is obviously singular when $s = 0$. However, this differential equation has the solution

$$x = c_1(1 - \cos s) + c_2 \sin s,$$

which is well defined.

Analytically speaking, the differential equation presents no problem since we already have the solution. This is not the case if we are solving the differential equation *numerically*:

$$x'' = f(x', x, s) = \frac{\sin s}{1 - \cos s} x' - \frac{x}{1 - \cos s};$$

in this case the singularity presents a problem of which we are only aware when the disaster occurs.

Stiefel and Scheifele [64] present the case of one-dimensional two-body motion as an example of the advantage of regularization. Suppose we have two masses acting under influence of gravity along the x-axis. The equation of motion of m_2 with respect to m_1 is

$$\ddot{x} + \frac{\mu}{x^2} = 0, \tag{9.1}$$

where $\mu = G(m_1 + m_2)$. This differential equation has the integral

$$h = \frac{1}{2}\dot{x}^2 - \frac{\mu}{x} = \text{constant}, \tag{9.2}$$

where h is the two-body (Keplerian) energy. Equations (9.1) and (9.2) are a special case of equations (2.8) and (2.16) of chapter 2. Equation (9.1) has a *singularity* at $x = 0$. From equation (9.2) we note that

$$\dot{x} \to \infty \quad \text{as} \quad x \to 0.$$

That is, the velocity ($\dot{x}$) becomes infinite at the origin.

Note if we transform from time t to "fictitious" time (s) by employing the Sundmann transformation

$$\frac{dt}{ds} = x, \tag{9.3}$$

then the "velocity" (x') does *not* become infinite since

$$x' = \frac{dx}{ds} = \frac{dx}{dt}\frac{dt}{ds} = x\frac{dx}{dt} = x\dot{x}. \tag{9.4}$$

Clearly, as $x \to 0$, $\dot{x} \to \infty$, but $x' \to 0$.

The *first step* in regularization is to change the independent variable from t to s in equations (9.1) and (9.2). Note from equation (9.4) that

$$\dot{x} = \frac{x'}{x}. \tag{9.5}$$

The second derivative is

$$\ddot{x} = \frac{d}{dt}(\dot{x}) = \frac{d}{dt}\left(\frac{x'}{x}\right) = \frac{d}{ds}\left(\frac{x'}{x}\right)\frac{ds}{dt}$$

$$\ddot{x} = \frac{x''}{x^2} - \frac{(x')^2}{x^3}. \tag{9.6}$$

Now substitute equation (9.6) into (9.1) to obtain

$$x'' - \frac{(x')^2}{x} = -\mu, \tag{9.7}$$

which is the differential equation of motion with s as the new independent variable. Now substitute equation (9.4) into (9.2) to obtain an expression for the energy involving x' instead of $\dot{x}$:

$$h = \frac{1}{2}\frac{(x')^2}{x^2} - \frac{\mu}{x}. \tag{9.8}$$

Note that the differential equation of motion (eq. 9.7) is still singular because of the $(x')^2/x$ factor. The differential equation of motion has not been regularized by changing from t to s.

The next step in regularizing equation (9.1) is suggested by noting that the $(x')^2/x$ term can be eliminated from equation (9.7) by using equation (9.8), which we rearrange slightly to get

$$\frac{(x')^2}{x} = 2\mu + 2hx. \tag{9.9}$$

Now substitute equation (9.9) into (9.7):

$$x'' - 2\mu - 2hx = -\mu$$

or

$$x'' - 2hx = \mu. \tag{9.10}$$

The differential equation of motion has now been regularized since x does not appear in the denominator. The process of eliminating a singularity from a differential equation of motion by the use of one or more integrals is called *embedding*.

For the rectilinear ellipse ($h < 0$), equation (9.10) has the solution

$$x = -\frac{\mu}{2h} + c_1 \cos \sqrt{-2h}\, s + c_2 \sin \sqrt{-2h}\, s;$$

c_1 and c_2 are related to the initial conditions x_0 and x_0'. The time equation is obtained from equation (9.3):

$$t = \int x\, ds = c_0 - \frac{\mu}{2h}s + \frac{c_1}{\sqrt{-2h}} \sin \sqrt{-2h}\, s - \frac{c_2}{\sqrt{-2h}} \cos \sqrt{-2h}\, s.$$

9.3 Regularizing the Two-Body Problem

This approach to the regularization of the two-body problem was published by Sperling [61]. The differential equation for the perturbed two-body problem is

$$\ddot{\mathbf{r}} + \frac{\mu}{r^3}\mathbf{r} = \mathbf{F}. \tag{9.11}$$

Note that the regularization will be applied to the two-body part of the problem *only*. The perturbation $\mathbf{F}$ is carried along for "bookkeeping" purposes for the time being.

The *first step* is to transform the independent variable from time to fictitious time by the Sundmann transformation,

$$\frac{dt}{ds} = r = \|\mathbf{r}\|. \tag{9.12}$$

The "velocity" and "acceleration" in the fictitious time are derived from $\mathbf{r} = \mathbf{r}(s(t))$, in a manner analogous to the previous section:

$$\dot{\mathbf{r}} = \frac{d\mathbf{r}}{ds}\frac{ds}{dt} = \frac{\mathbf{r}'}{r}. \tag{9.13}$$

Also,

$$\ddot{\mathbf{r}} = \frac{d\dot{\mathbf{r}}}{dt} = \frac{d}{ds}\left(\frac{\mathbf{r}'}{r}\right)\frac{ds}{dt} = \frac{1}{r^2}\mathbf{r}'' - \frac{r'\mathbf{r}'}{r^3}.$$

Using the relationship

$$r' = \frac{\mathbf{r} \cdot \mathbf{r}'}{r},$$

we get

$$\ddot{\mathbf{r}} = \frac{1}{r^2}\mathbf{r}'' - \frac{1}{r^4}(\mathbf{r} \cdot \mathbf{r}')\mathbf{r}'. \tag{9.14}$$

Now substitute equation (9.14) into (9.11) and multiply through by r^2 to get

$$\mathbf{r}'' - \frac{1}{r^2}(\mathbf{r} \cdot \mathbf{r}')\mathbf{r}' + \frac{\mu}{r}\mathbf{r} = r^2\mathbf{F}. \tag{9.15}$$

We call equation (9.15) "the equation of the perturbed two-body problem in the fictitious time." This is an important equation, used in several other approaches to regularization.

Compare equation (9.15) with equation (9.7) for the one-dimensional case where we completed the regularization by using the energy equation. We now try the same approach here.

The energy integral, a constant when perturbations are absent, is

$$h = \frac{1}{2}\dot{\mathbf{r}} \cdot \dot{\mathbf{r}} - \frac{\mu}{r}. \tag{9.16}$$

Using equation (9.13) to transform to fictitious time,

$$h = \frac{1}{2}\frac{\mathbf{r}' \cdot \mathbf{r}'}{r^2} - \frac{\mu}{r}. \tag{9.17}$$

Recall that in the one-dimensional case we used the energy to eliminate the "velocity," but for this case (the perturbed two-body problem) the "velocity" term in equation (9.15) *cannot* be eliminated by using equation (9.17). Following Sperling we employ the Laplace vector (a constant when perturbations are absent),

$$\mu\boldsymbol{\epsilon} = \dot{\mathbf{r}} \times \mathbf{c} - \frac{\mu}{r}\mathbf{r}$$

$$= \dot{\mathbf{r}} \times (\mathbf{r} \times \dot{\mathbf{r}}) - \frac{\mu}{r}\mathbf{r}. \tag{9.18}$$

Expanding the triple vector product, we obtain

$$\mu\boldsymbol{\epsilon} = [(\dot{\mathbf{r}} \cdot \dot{\mathbf{r}})\mathbf{r} - (\dot{\mathbf{r}} \cdot \mathbf{r})\dot{\mathbf{r}}] - \frac{\mu}{r}\mathbf{r}.$$

Again using equation (9.13) to transform to fictitious time,

$$\mu\boldsymbol{\epsilon} = \frac{1}{r^2}[(\mathbf{r}' \cdot \mathbf{r}')\mathbf{r} - (\mathbf{r}' \cdot \mathbf{r})\mathbf{r}'] - \frac{\mu}{r}\mathbf{r}. \tag{9.19}$$

Now we use equation (9.19) to eliminate the $(\mathbf{r}' \cdot \mathbf{r})\mathbf{r}'$ term from equation (9.15), obtaining

$$\mathbf{r}'' + \left[\mu\boldsymbol{\epsilon} - \frac{1}{r^2}(\mathbf{r}' \cdot \mathbf{r}')\mathbf{r} + \frac{\mu}{r}\mathbf{r}\right] + \frac{\mu}{r}\mathbf{r} = r^2\mathbf{F}.$$

Transfer the $\mu\epsilon$ term to the right side to get

$$\mathbf{r}'' + \left[\frac{2\mu}{r} - \frac{1}{r^2}(\mathbf{r}'\cdot\mathbf{r}')\right]\mathbf{r} = -\mu\epsilon + r^2\mathbf{F}. \tag{9.20}$$

Comparing equation (9.20) with equation (9.17) we see that the bracketed term in (9.20) is $-2h$. Therefore equation (9.20) becomes

$$\mathbf{r}'' + \alpha_k\mathbf{r} = -\mu\epsilon + r^2\mathbf{F}, \tag{9.21}$$

where $\alpha_k = -2h$.

Equation (9.11) has been *linearized* and *regularized* (when $\mathbf{F} = 0$) by changing the independent variable from t to s using the Sundmann transformation (eq. 9.12) and by embedding the Laplacian integral $\mu\epsilon$ and the energy h. The result is equation (9.21), which we may also use to give us a linearized and regularized differential equation for the distance. Take the dot product of equation (9.21) with $\mathbf{r}$,

$$\mathbf{r}\cdot\mathbf{r}'' + \alpha_k\mathbf{r}\cdot\mathbf{r} = -\mu\mathbf{r}\cdot\mathbf{\epsilon} + r^2\mathbf{r}\cdot\mathbf{F}. \tag{9.22}$$

Since

$$r' = \frac{\mathbf{r}\cdot\mathbf{r}'}{r},$$

we take the derivative to get

$$r'' = -\frac{1}{r^2}r'(\mathbf{r}\cdot\mathbf{r}') + \frac{1}{r}(\mathbf{r}\cdot\mathbf{r}'' + \mathbf{r}'\cdot\mathbf{r}').$$

Solving for $\mathbf{r}\cdot\mathbf{r}''$,

$$\mathbf{r}\cdot\mathbf{r}'' = rr'' - \mathbf{r}'\cdot\mathbf{r}' + \frac{1}{r}r'(\mathbf{r}\cdot\mathbf{r}'). \tag{9.23}$$

Now substitute equation (9.19) for $\mu\epsilon$ on the right side of equation (9.22) and use equation (9.23) on the left side of equation (9.22) to eliminate $\mathbf{r}\cdot\mathbf{r}''$:

$$rr'' - \mathbf{r}'\cdot\mathbf{r}' + \frac{1}{r}r'(\mathbf{r}\cdot\mathbf{r}') + \alpha_k r^2$$

$$= -\frac{1}{r^2}\left[(\mathbf{r}'\cdot\mathbf{r}')r^2 - (\mathbf{r}'\cdot\mathbf{r})(\mathbf{r}'\cdot\mathbf{r})\right] + \frac{\mu}{r}\mathbf{r}\cdot\mathbf{r} + r^2\mathbf{r}\cdot\mathbf{F}.$$

After canceling and dividing by r, this reduces to

$$r'' + \alpha_k r = \mu + r\mathbf{r}\cdot\mathbf{F}. \tag{9.24}$$

Compare equation (9.24) with equation (9.10) of the one-dimensional problem. When $\mathbf{F} = 0$, equation (9.24) has an oscillator-type solution for r with a constant term. This solution for r can now be integrated for t.

Table 9.1   The two-body constants.

No.	Equation	Two-Body Constants
(9.21)	$r'' + \alpha_k r = -\mu\varepsilon + r^2 \mathbf{F}$	$\mathbf{r}_0, \mathbf{r}'_0 \to$ 6 constants
(9.24)	$r'' + \alpha_k r = \mu + r\mathbf{r} \cdot \mathbf{F}$	$r_0, r'_0 \to$ 2 constants
(9.12)	$t' = r$	$t_0 \to$ 1 constant
(9.25)	$\alpha_k' = -2\mathbf{r}' \cdot \mathbf{F}$	$h \to$ 1 constant
(9.26)	$\mu\varepsilon' = 2(\mathbf{r}' \cdot \mathbf{F})\mathbf{r} - (\mathbf{r} \cdot \mathbf{F})\mathbf{r}' - (\mathbf{r}' \cdot \mathbf{r})\mathbf{F}$	$\varepsilon \to$ 3 constants

The differential equation for the Keplerian energy is found from equation (9.17) (recalling that $\alpha_k = -2h$) by employing Poisson's method,

$$\alpha_k' = \frac{\partial \alpha_k}{\partial \mathbf{r}'} \cdot (r^2\mathbf{F}). \tag{9.25}$$

The differential equation for the Laplace vector is found from equation (9.19) by employing Poisson's method,

$$\mu\varepsilon' = \frac{\partial \mu\varepsilon}{\partial \mathbf{r}'} (r^2\mathbf{F}). \tag{9.26}$$

From these equations it is clear that a minimum of thirteen constants of integration are required (see table 9.1). There are of course only six constants, but we have introduced redundant variables in order to *regularize* the differential equation (9.11). These thirteen constants will become the new "variables" when we apply the Variation of Parameters technique to the entire problem. The development of differential equations for these constants, or for constants related to them was first done by Burdet [25]. We note that Burdet used the Keplerian energy as a variable. A modified version of Burdet's method was produced in 1973 by Bond and Hanssen [19], who used the total energy as a variable. Burdet's method was again modified in 1989 by Bond and Gottlieb [18] who used the Jacobi integral (eq. 8.61) as defined in chapter 8 instead of the Keplerian energy.

In general we can define without loss of generality the perturbation $\mathbf{F}$ to be of the form

$$\mathbf{F} = \mathbf{P} - \frac{\partial V(\mathbf{r}, t)}{\partial \mathbf{r}}.$$

The $\mathbf{P}$ components are perturbations *not* generally derivable from a potential. For example, thrust and drag must be included in $\mathbf{P}$. $V(\mathbf{r}, t)$ is the perturbing potential that can depend explicitly on time if required. A special case of this potential is of the form $V(\mathbf{r})$, which depends on the position $\mathbf{r}$ only. The perturbing potential is always due to gravitational effects, such as an irregularly shaped rotating planet or a perturbing third body.

9.3.1 The Jacobi Integral

Recall that in §8.7.3 we showed for the case where

$$\ddot{\mathbf{r}} + \frac{\mu}{r^3}\mathbf{r} = -\frac{\partial V(\mathbf{r}, t)}{\partial \mathbf{r}}$$

that an integral of the motion could be determined for the case where

$$\frac{\partial V(\mathbf{r}, t)}{\partial t} = -\boldsymbol{\omega} \cdot \mathbf{r} \times \frac{\partial V(\mathbf{r}, t)}{\partial \mathbf{r}},$$

where $\boldsymbol{\omega}(=$ a constant vector). We found a constant of the motion to be

$$h + V(\mathbf{r}, t) - \boldsymbol{\omega} \cdot (\mathbf{r} \times \dot{\mathbf{r}}) = \text{constant} = -\frac{\alpha_J}{2},$$

where now we define α_J to be the Jacobi integral. Since $\alpha_k = -2h$, the integral becomes

$$-\frac{\alpha_k}{2} + V(\mathbf{r}, t) - \boldsymbol{\omega} \cdot (\mathbf{r} \times \dot{\mathbf{r}}) = -\frac{\alpha_J}{2}.$$

Solve for α_k:

$$\alpha_k = \alpha_J + 2V(\mathbf{r}, t) - 2\boldsymbol{\omega} \cdot \mathbf{c},$$

where $\mathbf{c} = \mathbf{r} \times \dot{\mathbf{r}}$.

9.3.2 Change to the Jacobi Integral

Recall equation (9.21):

$$\mathbf{r}'' + \alpha_k \mathbf{r} = -\mu\boldsymbol{\varepsilon} + r^2\mathbf{F}.$$

Substitute the solution from the last section for α_k:

$$\mathbf{r}'' + (\alpha_J - 2\boldsymbol{\omega} \cdot \mathbf{c} + 2V(\mathbf{r}, t))\mathbf{r} = -\mu\boldsymbol{\varepsilon} + r^2\mathbf{F}.$$

Move the perturbing potential V to the right side, and define the axial element,

$$\sigma = \boldsymbol{\omega} \cdot \mathbf{c}, \tag{9.27}$$

which we also consider a perturbation and move to the right side. We have, therefore,

$$\mathbf{r}'' + \alpha_J \mathbf{r} = -\mu\boldsymbol{\varepsilon} + r^2\mathbf{F} + 2(\sigma - V(\mathbf{r}, t))\mathbf{r}. \tag{9.28}$$

Similarly, equation (9.24) for the distance becomes

$$r'' + \alpha_J r = \mu + r\mathbf{r} \cdot \mathbf{F} + 2(\sigma - V(\mathbf{r}, t))r. \tag{9.29}$$

Although σ is of the order of the perturbation, it is a new element having a differential equation. Finally, note that the definition of the α_J becomes

$$\alpha_J = \alpha_k - 2V(\mathbf{r}, t) + 2\sigma. \tag{9.30}$$

9.3.3 The Two-Body Solution

In the absence of a perturbation, that is, $\mathbf{F} = 0$ and $\boldsymbol{\omega} = 0$, equation (9.28) and (9.29) become

$$\mathbf{r}'' + \alpha_J \mathbf{r} = -\mu\boldsymbol{\varepsilon} \tag{9.31}$$

$$r'' + \alpha_J r = \mu. \tag{9.32}$$

Note that the elements $\boldsymbol{\varepsilon}$, $\alpha_J (= \alpha_k)$ and σ are all *constant*.

We write the solution to equation (9.31) in terms of the Stumpff functions c_0, c_1, and c_2[1] and the initial conditions (at $s = 0$) $\mathbf{r}_0$ and $\mathbf{r}'_0$,

$$\mathbf{r} = \mathbf{r}_0 c_0 + \mathbf{r}'_0 s c_1 - \mu\boldsymbol{\varepsilon} s^2 c_2. \tag{9.33}$$

We now differentiate equation (9.33) twice and use Appendix E to prove that it is a solution to the differential equation (9.31). Differentiating equation (9.33),

$$\mathbf{r}' = \mathbf{r}_0 c'_0 + \mathbf{r}'_0 \left(s c'_1 + c_1\right) - \mu\boldsymbol{\varepsilon} \left(s^2 c'_2 + 2s c_2\right). \tag{9.34}$$

From Appendix E we have

equation (E.5) with $n = 0$: $c'_0 = -\alpha_J s c_1$
equation (E.4) with $n = 1$: $s c'_1 = c_0 - c_1$
equation (E.4) with $n = 2$: $s c'_2 = c_1 - 2c_2$.

Using these results, equation (9.34) becomes

$$\mathbf{r}' = -(\alpha_J \mathbf{r}_0 + \mu\boldsymbol{\varepsilon}) s c_1 + \mathbf{r}'_0 c_0, \tag{9.35}$$

which is the velocity in the fictitious time. Differentiate equation (9.35) to

---

[1]See Appendix E for an explanation of Stumpff functions. Note that the arguments of the Stumpff functions here are implied; that is, $c_k = c_k(\alpha_J s^2)$.

obtain

$$\mathbf{r}'' = -(\alpha_{,}\mathbf{r}_0 + \mu\varepsilon)(sc_1' + c_1) + \mathbf{r}_0'c_0'.$$

Again using equation (E.4) we obtain

$$\mathbf{r}'' = -\alpha_{,}(\mathbf{r}_0c_0 + \mathbf{r}_0'sc_1) - \mu\varepsilon c_0. \tag{9.36}$$

From equation (E.1) with $z = \alpha_{,}s^2$ and $n = 0$,

$$c_0\left(\alpha_{,}s^2\right) + \alpha_{,}s^2c_2\left(\alpha_{,}s^2\right) = 1. \tag{9.37}$$

Solve equation (9.37) for $c_0(\alpha_{,}s^2)$ and substitute into equation (9.36) to obtain

$$\mathbf{r}'' = -\alpha_{,}\left(\mathbf{r}_0c_0 + \mathbf{r}_0'sc_1\right) - \mu\varepsilon\left(1 - \alpha_{,}s^2c_2\right)$$
$$= -\alpha_{,}\left(\mathbf{r}_0c_0 + \mathbf{r}_0'sc_1 - \mu\varepsilon s^2c_2\right) - \mu\varepsilon.$$

Since the coefficient of $\alpha_{,}$ is the position $\mathbf{r}$, equation (9.33), we can write the last equation as

$$\mathbf{r}'' + \alpha_{,}\mathbf{r} = -\mu\varepsilon,$$

which is equation (9.31). This verifies that equation (9.33) is indeed the solution of the differential equation (9.31).

Since the differential equation (9.32) for the scalar distance r is similar (replace $\mathbf{r}$ by r and $-\mu\varepsilon$ by μ) to equation (9.31), the solution for equation (9.32) is

$$r = r_0c_0 + r_0'sc_1 + \mu s^2c_2, \tag{9.38}$$

with the radial velocity

$$r' = (\mu - \alpha_{,}r_0)sc_1 + r_0'c_0, \tag{9.39}$$

where r_0 and r_0' are the initial conditions of r and r'. Using the Sundmann transformation (eq. 9.12),

$$\frac{dt}{ds} = r,$$

we can obtain an equation for the time by using equation (9.38) for r,

$$t = t_0 + r_0 \int c_0 ds + r_0' \int sc_1 ds + \mu \int s^2c_2 ds. \tag{9.40}$$

Using equations (E.7), (E.8), and (E.9), equation (9.40) becomes

$$t = t_0 + r_0 s c_1 + r_0' s^2 c_2 + \mu s^3 c_3. \tag{9.41}$$

Note that $s = 0$ when $t = t_0$. Equation (9.41) is Kepler's equation written in terms of the Stumpff functions.

The *two-body* solution is therefore equations (9.33) and (9.35) for $\mathbf{r}$ and $\mathbf{r}'$, and equation (9.41) for t. Equations (9.38) and (9.39) for r and r' are optional in the two-body solution but will be useful in the *perturbed two-body* VOP procedure.

9.3.4 Introduction of δ and γ

Here we develop different forms for the position and velocity in the fictitious time. We substitute

$$c_0 + \alpha_J s^2 c_2 = 1$$

(which is obtained from equation E.1 with $n = 0$) into equation (9.33) to get

$$\mathbf{r} = \mathbf{r}_0 \left(1 - \alpha_J s^2 c_2\right) + \mathbf{r}_0' s c_1 - \mu \boldsymbol{\varepsilon} s^2 c_2.$$

Since the Laplace vector becomes undefined for the case where the eccentricity becomes zero, we replace it with another integral of the motion, which is defined by the relation

$$\boldsymbol{\delta} = -(\alpha_J \mathbf{r}_0 + \mu \boldsymbol{\varepsilon}), \tag{9.42}$$

and the position $\mathbf{r}$ becomes

$$\mathbf{r} = \mathbf{r}_0 + \mathbf{r}_0' s c_1 + \boldsymbol{\delta} s^2 c_2. \tag{9.43}$$

Differentiate to get

$$\mathbf{r}' = \boldsymbol{\delta} \left(s^2 c_2' + 2 s c_2\right) + \mathbf{r}_0' \left(s c_1' + c_1\right),$$

or using equation (E.4) we obtain for the velocity in fictitious time,

$$\mathbf{r}' = \mathbf{r}_0' c_0 + \boldsymbol{\delta} s c_1. \tag{9.44}$$

In a similar manner we rewrite the scalar equation (9.38) for distance using equation (E.1). We obtain

$$r = r_0 + (-r_0 \alpha_J + \mu)s^2 c_2 + r_0' s c_1.$$

Now *define* the new element

$$\gamma = \mu - r_0 \alpha_J \tag{9.45}$$

and substitute into the distance equation to obtain

$$r = r_0 + r_0' s c_1 + \gamma s^2 c_2. \tag{9.46}$$

Differentiate to get

$$r' = r_0' c_0 + \gamma s c_1. \tag{9.47}$$

We can also develop another equation for the time. Since

$$\frac{dt}{ds} = r,$$

we can integrate equation (9.46),

$$dt = \left(r_0 + r_0' s c_1 + \gamma s^2 c_2 \right) ds,$$

to obtain

$$t = t_0 + r_0 s + r_0' s^2 c_2 + \gamma s^3 c_3. \tag{9.48}$$

Substitute equation (E.1) with $n = 1$ to finally obtain

$$t = t_0 + r_0 s c_1 + r_0' s^2 c_2 + \mu s^3 c_3,$$

which is the same as equation (9.41).

9.3.5 The Significance of δ and γ

Recall equation (9.31):

$$\mathbf{r}'' + \alpha_J \mathbf{r} = -\mu \boldsymbol{\varepsilon}.$$

Now define

$$\mathbf{x} = -\alpha_J \mathbf{r} - \mu \boldsymbol{\varepsilon},$$

which when substituted into equation (9.31) yields the differential equation

$$\mathbf{x}'' + \alpha_J \mathbf{x} = 0.$$

We have transformed equation (9.31) to an oscillator *without the constant* on the right side. The solution to the differential equation is

$$\mathbf{x} = \mathbf{x}_0 c_0 + \mathbf{x}_0' s c_1,$$

as can be easily verified by performing the differentiation and substitution. Note that when $s = 0$,

$$\mathbf{x}(0) = \mathbf{x}_0 = -\alpha_J \mathbf{r}_0 - \mu\varepsilon \equiv \boldsymbol{\delta}.$$

So the new element $\boldsymbol{\delta}$ is the initial value of $\mathbf{x}$.

Similarly, for the distance equation (9.32),

$$r'' + \alpha_J r = \mu,$$

we define

$$y = \mu - \alpha_J r.$$

Substitute into equation (9.32) to obtain

$$y'' + \alpha_J y = 0,$$

which has the solution

$$y = y_0 c_0 + y_0' s c_1.$$

When $s = 0$ we get

$$y(0) = y_0 = \mu - \alpha_J r_0 \equiv \gamma.$$

So the new element γ is the initial value of y.

9.4 Summary of the Elements

The elements that will be treated by the Variation of Parameters method are as follows:

1. The *initial conditions* associated with the spatial differential equations (9.31) are $\mathbf{r}_0$ and $\mathbf{r}_0'$. Also associated with equation (9.31) is the new element (eq. 9.42),

$$\boldsymbol{\delta} = -\alpha_J \mathbf{r}_0 - \mu\varepsilon.$$

The elements r_0, r_0', and $\boldsymbol{\delta}$ are called the *spatial* elements since the solution of equation (9.31) will define the orbit in space but *not* in time. Note that the Laplace vector ($\mu\boldsymbol{\varepsilon}$) has been replaced by $\boldsymbol{\delta}$.

2. The *initial conditions* associated with the temporal differential equations (9.32) and (9.12),

$$r'' + \alpha_J r = \mu$$

$$t' = r.$$

These initial conditions are r_0, r_0', and t_0. Also associated with equation (9.32) is the new element defined by equation (9.45),

$$\gamma = \mu - r_0 \alpha_J.$$

The elements r_0, r_0', t_0, and γ are called the *temporal* elements since the solution of equations (9.12) and (9.32) define the orbit in time.

3. The other two elements are the Jacobian and axial elements α_J and σ, which are defined by equations (9.27) and (9.30),

$$\sigma = \boldsymbol{\omega} \cdot \mathbf{c}$$

$$\alpha_J = \alpha_k - 2V(\mathbf{r}, t) - 2\sigma,$$

where $\mathbf{c} = \mathbf{r} \times \dot{\mathbf{r}}$ and $\alpha_k = -2h$. Note that σ is *not* a two-body element and will appear only in the perturbation. Also note that α_J *is* a two-body element since in the absence of perturbations, $\alpha_J = \alpha_k = constant$.

9.5 Sperling-Burdet Approach

In the remainder of this chapter we will follow the approach introduced by Bond and Fraietta [17] and treat the *spatial* and *temporal* equations separately in the Variation of Parameters (VOP) technique when perturbations are introduced. For the perturbed case, we will first develop the differential equations for σ and α_J by the "direct" VOP method. We will then use the Lagrange VOP method to develop differential equations for the spatial elements, and finally use the Lagrange VOP method to develop differential equations for the temporal elements. This approach is a further modification of Burdet's method.

First we will develop the differential equations for the axial and Jacobian elements. Recall equation (9.30) for the Jacobian,

$$\alpha_J = \alpha_k + 2\sigma - 2V, \tag{9.49}$$

where $V = V(\mathbf{r}, t)$, and also equation (9.27) for the axial element,

$$\sigma = \boldsymbol{\omega} \cdot \mathbf{c} = \boldsymbol{\omega} \cdot (\mathbf{r} \times \dot{\mathbf{r}}). \qquad (9.50)$$

We first derive $\dot{\sigma}$, then use the Sundmann transformation to formulate σ'. Differentiating equation (9.50) with respect to time yields

$$\dot{\sigma} = \boldsymbol{\omega} \cdot (\dot{\mathbf{r}} \times \dot{\mathbf{r}}) + \boldsymbol{\omega} \cdot (\mathbf{r} \times \ddot{\mathbf{r}}).$$

Solve equation (9.11) for $\ddot{\mathbf{r}}$ and substitute into the above equation, and simplify using $\mathbf{r} \times \mathbf{r} = 0$ and $\dot{\mathbf{r}} \times \dot{\mathbf{r}} = 0$ to get

$$\dot{\sigma} = \boldsymbol{\omega} \cdot \mathbf{r} \times \mathbf{F}. \qquad (9.51)$$

Since from Sundmann's transformation,

$$\dot{\sigma} = \frac{d\sigma}{ds}\frac{ds}{dt} = \frac{1}{r}\sigma',$$

we have

$$\sigma' = r\boldsymbol{\omega} \cdot \mathbf{r} \times \dot{\mathbf{F}}. \qquad (9.52)$$

Similarly we first derive $\dot{\alpha}_J$ and then α'_J. Differentiating equation (9.49),

$$\dot{\alpha}_J = \dot{\alpha}_k + 2\dot{\sigma} - 2\frac{dV}{dt}. \qquad (9.53)$$

Since $\alpha_k = -2h$, we have, using equation (8.37),

$$\dot{\alpha}_k = -2\dot{h} = -2\dot{\mathbf{r}} \cdot \mathbf{F}. \qquad (9.54)$$

Using equations (9.51) and (9.54) in equation (9.53), we get

$$\dot{\alpha}_J = -2\dot{\mathbf{r}} \cdot \mathbf{F} + 2\boldsymbol{\omega} \cdot \mathbf{r} \times \mathbf{F} - 2\frac{dV}{dt}.$$

Recall that we earlier defined the perturbation to be of the form

$$\mathbf{F} = \mathbf{P} - \frac{\partial V}{\partial \mathbf{r}},$$

and since the total derivative of V is

$$\frac{dV}{dt} = \frac{\partial V}{\partial \mathbf{r}} \cdot \dot{\mathbf{r}} + \frac{\partial V}{\partial t},$$

the equation for $\dot{\alpha}_J$ becomes

$$\dot{\alpha}_J = -2\dot{\mathbf{r}} \cdot \left(\mathbf{P} - \frac{\partial V}{\partial \mathbf{r}}\right) + 2\boldsymbol{\omega} \cdot \mathbf{r} \times \left(\mathbf{P} - \frac{\partial V}{\partial \mathbf{r}}\right) - 2\frac{\partial V}{\partial \mathbf{r}} \cdot \dot{\mathbf{r}} - 2\frac{\partial V}{\partial t}.$$

Note the cancelation and now recall equation (8.57) since we are considering time-dependent potentials for which

$$\frac{\partial V}{\partial t} = -\boldsymbol{\omega} \cdot \mathbf{r} \times \frac{\partial V}{\partial \mathbf{r}}.$$

Substitute into the equation for $\dot{\alpha}_J$ and cancel terms to obtain

$$\dot{\alpha}_J = 2(-\dot{\mathbf{r}} + \boldsymbol{\omega} \times \mathbf{r}) \cdot \mathbf{P}. \tag{9.55}$$

Since from Sundmann's transformation

$$\dot{\alpha}_J = \frac{d(\alpha_J)}{ds} \frac{ds}{dt} = \frac{1}{r} \alpha_J'$$

and

$$\dot{\mathbf{r}} = \frac{1}{r} \mathbf{r}',$$

we get finally

$$\alpha_J' = 2(-\mathbf{r}' + r\,\boldsymbol{\omega} \times \mathbf{r}) \cdot \mathbf{P}. \tag{9.56}$$

Note that when $\mathbf{P} = 0$, α_J is constant, and when $\boldsymbol{\omega} = 0$, σ is constant.

9.5.1 Equations for the Spatial Elements

We write the solution for the *spatial* system as

$$\mathbf{x}_1 = \mathbf{r} = \boldsymbol{\alpha} + \boldsymbol{\beta} s c_1 + \boldsymbol{\delta} s^2 c_2 \tag{9.57}$$

$$\mathbf{x}_2 = \mathbf{r}' = \boldsymbol{\beta} c_0 + \boldsymbol{\delta} s c_1 \tag{9.58}$$

$$\mathbf{x}_3 = -\alpha_J \mathbf{r} - \mu \boldsymbol{\varepsilon} = -x_5 \mathbf{x}_1 - \mu \boldsymbol{\varepsilon} \tag{9.59}$$

$$x_4 = \sigma \tag{9.60}$$

$$x_5 = \alpha_J, \tag{9.61}$$

where we introduce the notation

$$\boldsymbol{\alpha} = \mathbf{r}_0 = \mathbf{x}_1(0)$$

$$\boldsymbol{\beta} = \mathbf{r}_0' = \mathbf{x}_2(0)$$

$$\boldsymbol{\delta} = \mathbf{x}_3(0) = -\alpha_J \boldsymbol{\alpha} - \mu \boldsymbol{\varepsilon}.$$

We have already derived a differential equation for α_J (eq. 9.56), but we must also include it here once more since it appears in the Stumpff functions,

$$c_k = c_k \left(\alpha_J s^2 \right).$$

We also have derived a differential equation for σ (eq. 9.52). It is not necessary to include it, but we carry it through for completeness.

Now, from equation (9.28) for the *perturbed spatial* system, we will derive a first-order system of differential equations of the form

$$\mathbf{x}' = \mathbf{H} + \mathbf{G},$$

where

$$\mathbf{x}^T = (\mathbf{x}_1, \mathbf{x}_2, \mathbf{x}_3, x_4, x_5),$$
$$\mathbf{H} = \text{the unperturbed part, and}$$
$$\mathbf{G} = \text{the perturbation.}$$

Now from the definition implied by equations (9.57)–(9.61) and equation (9.28), we derive

$$\mathbf{x}_1' = \mathbf{r}' = \mathbf{x}_2$$
$$\mathbf{x}_2' = \mathbf{r}'' = -\alpha_J \mathbf{r} - \mu \boldsymbol{\varepsilon} + r^2 \mathbf{F} + 2(\sigma - V)\mathbf{r}$$
$$\quad = \mathbf{x}_3 + r^2 \mathbf{F} + 2(\sigma - V)\mathbf{r}$$
$$\mathbf{x}_3' = -\alpha_J \mathbf{r}' - \alpha_J' \mathbf{r} - \mu \boldsymbol{\varepsilon}'$$
$$\quad = -x_5 \mathbf{x}_2 - (\alpha_J' \mathbf{r} + \mu \boldsymbol{\varepsilon}')$$
$$x_4' = \sigma'$$
$$x_5' = \alpha_J'.$$

We put these derivatives in the necessary form:

$$\mathbf{x}_1' = \mathbf{x}_2 + \mathbf{G}_1 \tag{9.62}$$
$$\mathbf{x}_2' = \mathbf{x}_3 + \mathbf{G}_2 \tag{9.63}$$
$$\mathbf{x}_3' = -x_5 \mathbf{x}_2 + \mathbf{G}_3 \tag{9.64}$$
$$x_4' = g_4 \tag{9.65}$$
$$x_5' = g_5, \tag{9.66}$$

where we have defined the perturbations

$$\mathbf{G}_1 = 0 \tag{9.67}$$

$$\mathbf{G}_2 = r^2\mathbf{F} + 2(\sigma - V)\mathbf{r} \tag{9.68}$$

$$\mathbf{G}_3 = -\alpha'_J\mathbf{r} - \mu\boldsymbol{\varepsilon}' \tag{9.69}$$

$$g_4 = r\boldsymbol{\omega} \cdot \mathbf{r} \times \mathbf{F} \tag{9.70}$$

$$g_5 = 2(-\mathbf{r}' + r\boldsymbol{\omega} \times \mathbf{r}) \cdot \mathbf{P}. \tag{9.71}$$

Note that in previous work (Bond and Gottlieb [18], eq. 2.17) that $\mathbf{Q} \equiv \mathbf{G}_2$. Also, we have not introduced the x-notation in $\mathbf{G}$ since they are perturbations.

In Lagrange's method the differential equation for the "constants" (which are now the new variables) is given by equation (8.32),

$$\frac{\partial \mathbf{x}}{\partial \mathbf{c}}\mathbf{c}' = \mathbf{G},$$

where here we define

$$\mathbf{c}^T = \left(\alpha^T, \boldsymbol{\beta}^T, \boldsymbol{\delta}^T, \sigma, \alpha_J\right).$$

We obtain the matrix of partials from the solution given by equations (9.57)–(9.61), where α, $\boldsymbol{\beta}$, $\boldsymbol{\delta}$, σ, and α_J are functions of s.

For $\mathbf{x}_1$ we have

$$\mathbf{x}_1 = \mathbf{r} = \alpha + \boldsymbol{\beta}sc_1 + \boldsymbol{\delta}s^2c_2.$$

Taking the partials,

$$\frac{\partial \mathbf{x}_1}{\partial \mathbf{c}} = \left[\begin{array}{ccccc} \dfrac{\partial \mathbf{r}}{\partial \alpha} & \dfrac{\partial \mathbf{r}}{\partial \boldsymbol{\beta}} & \dfrac{\partial \mathbf{r}}{\partial \boldsymbol{\delta}} & \dfrac{\partial \mathbf{r}}{\partial \sigma} & \dfrac{\partial \mathbf{r}}{\partial \alpha_J} \end{array}\right],$$

which become

$$\frac{\partial \mathbf{x}_1}{\partial \mathbf{c}} = \left[\begin{array}{ccccc} I & Isc_1 & Is^2c_2 & \mathbf{0} & \dfrac{\partial \mathbf{r}}{\partial \alpha_J} \end{array}\right], \tag{9.72}$$

where I is the 3×3 identity matrix and $\mathbf{0}$ is a null column vector.

For $\mathbf{x}_2$ we have

$$\mathbf{x}_2 = \mathbf{r}' = \boldsymbol{\beta}c_0 + \boldsymbol{\delta}sc_1.$$

Taking the partials,

$$\frac{\partial \mathbf{x}_2}{\partial \mathbf{c}} = \left[\frac{\partial \mathbf{r}'}{\partial \boldsymbol{\alpha}} \quad \frac{\partial \mathbf{r}'}{\partial \boldsymbol{\beta}} \quad \frac{\partial \mathbf{r}'}{\partial \boldsymbol{\delta}} \quad \frac{\partial \mathbf{r}'}{\partial \sigma} \quad \frac{\partial \mathbf{r}'}{\partial \alpha_J} \right]$$

which become

$$\frac{\partial \mathbf{x}_2}{\partial \mathbf{c}} = \left[[0] \quad I c_0 \quad I s c_1 \quad \mathbf{0} \quad \frac{\partial \mathbf{r}'}{\partial \alpha_J} \right], \tag{9.73}$$

where $[0]$ is the 3×3 null matrix.

For $\mathbf{x}_3$ we have

$$\mathbf{x}_3 = -\alpha_J \mathbf{r} - \mu \boldsymbol{\varepsilon},$$

but using $\boldsymbol{\delta} = -\alpha_J \boldsymbol{\alpha} - \mu \boldsymbol{\varepsilon}$, we can eliminate $\mu \boldsymbol{\varepsilon}$ from $\mathbf{x}_3$,

$$\mathbf{x}_3 = -\alpha_J \mathbf{r} + \boldsymbol{\delta} + \alpha_J \boldsymbol{\alpha},$$

so we have the following alternate for $\mathbf{x}_3$:

$$\mathbf{x}_3 = \alpha_J (\boldsymbol{\alpha} - \mathbf{r}) + \boldsymbol{\delta}.$$

Taking the partials,

$$\frac{\partial \mathbf{x}_3}{\partial \mathbf{c}} = \left[\frac{\partial \mathbf{x}_3}{\partial \boldsymbol{\alpha}} \quad \frac{\partial \mathbf{x}_3}{\partial \boldsymbol{\beta}} \quad \frac{\partial \mathbf{x}_3}{\partial \boldsymbol{\delta}} \quad \frac{\partial \mathbf{x}_3}{\partial \sigma} \quad \frac{\partial \mathbf{x}_3}{\partial \alpha_J} \right].$$

Using the above equation for $\mathbf{x}_3$, we get

$$\frac{\partial \mathbf{x}_3}{\partial \boldsymbol{\alpha}} = \alpha_J \left[\frac{\partial \boldsymbol{\alpha}}{\partial \boldsymbol{\alpha}} - \frac{\partial \mathbf{r}}{\partial \boldsymbol{\alpha}} \right] = \alpha_J [I - I] = [0]$$

as well as

$$\frac{\partial \mathbf{x}_3}{\partial \boldsymbol{\beta}} = -\alpha_J \frac{\partial \mathbf{r}}{\partial \boldsymbol{\beta}} = -I \alpha_J s c_1$$

and also

$$\frac{\partial \mathbf{x}_3}{\partial \boldsymbol{\delta}} = -\alpha_J \frac{\partial \mathbf{r}}{\partial \boldsymbol{\delta}} + I = -\alpha_J s^2 c_2 I + I = I \left(1 - \alpha_J s^2 c_2 \right) = I c_0.$$

Therefore,

$$\frac{\partial \mathbf{x}_3}{\partial \mathbf{c}} = \left[[0] \quad -I \alpha_J s c_1 \quad I c_0 \quad \mathbf{0} \quad \frac{\partial \mathbf{x}_3}{\partial \alpha_J} \right]. \tag{9.74}$$

We also have

$$x_4 = \sigma$$

and

$$x_5 = \alpha_J.$$

Taking the partials, we get

$$\frac{\partial x_4}{\partial \mathbf{c}} = [\,\mathbf{0}^T \quad \mathbf{0}^T \quad \mathbf{0}^T \quad 1 \quad 0\,] \tag{9.75}$$

and

$$\frac{\partial x_5}{\partial \mathbf{c}} = [\,\mathbf{0}^T \quad \mathbf{0}^T \quad \mathbf{0}^T \quad 0 \quad 1\,], \tag{9.76}$$

where $\mathbf{0}^T$ is a null row vector.

Substituting equations (9.72)–(9.76) into equation (8.32), we get

$$\begin{pmatrix} I & Isc_1 & Is^2c_2 & 0 & \frac{\partial \mathbf{r}}{\partial \alpha_J} \\ [0] & Ic_0 & Isc_1 & 0 & \frac{\partial \mathbf{r}'}{\partial \alpha_J} \\ [0] & -I\alpha_J sc_1 & Ic_0 & 0 & \frac{\partial \mathbf{x}_3}{\partial \alpha_J} \\ \mathbf{0}^T & \mathbf{0}^T & \mathbf{0}^T & 1 & 0 \\ \mathbf{0}^T & \mathbf{0}^T & \mathbf{0}^T & 0 & 1 \end{pmatrix} \begin{pmatrix} \boldsymbol{\alpha}' \\ \boldsymbol{\beta}' \\ \boldsymbol{\delta}' \\ \sigma' \\ \alpha_J' \end{pmatrix} = \begin{pmatrix} \mathbf{G}_1 \\ \mathbf{G}_2 \\ \mathbf{G}_3 \\ G_4 \\ G_5 \end{pmatrix} \tag{9.77}$$

which yield the equations

$$\boldsymbol{\alpha}' + \boldsymbol{\beta}'sc_1 + \boldsymbol{\delta}'s^2c_2 + \frac{\partial \mathbf{r}}{\partial \alpha_J}\alpha_J' = \mathbf{G}_1 = \mathbf{0} \tag{9.78}$$

$$\boldsymbol{\beta}'c_0 + \boldsymbol{\delta}'sc_1 + \frac{\partial \mathbf{r}'}{\partial \alpha_J}\alpha_J' = \mathbf{G}_2 = \mathbf{Q} \tag{9.79}$$

$$-\boldsymbol{\beta}'\alpha_J sc_1 + \boldsymbol{\delta}'c_0 + \frac{\partial \mathbf{x}_3}{\partial \alpha_J}\alpha_J' = \mathbf{G}_3 = -\alpha_J'\mathbf{r} - \mu\boldsymbol{\varepsilon}' \tag{9.80}$$

$$r\boldsymbol{\omega} \cdot \mathbf{r} \times \mathbf{F} = g_4 = \sigma' \tag{9.81}$$

$$2(-\mathbf{r}' + r\boldsymbol{\omega} \times \mathbf{r}) \cdot \mathbf{P} = g_5 = \alpha_J'. \tag{9.82}$$

Equation (9.80) can be simplified by using the alternate for $\mathbf{x}_3$ derived above, that is, $\mathbf{x}_3 = \alpha_J(\boldsymbol{\alpha} - \mathbf{r}) + \boldsymbol{\delta}$, to get the partial derivative $\frac{\partial \mathbf{x}_3}{\partial \alpha_J}$. Equation (9.80) becomes

$$-\boldsymbol{\beta}'\alpha_J sc_1 + \boldsymbol{\delta}'c_0 + \left(\boldsymbol{\alpha} - \mathbf{r} - \alpha_J\frac{\partial \mathbf{r}}{\partial \alpha_J}\right)\alpha_J' = -\alpha_J'\mathbf{r} - \mu\boldsymbol{\varepsilon}', \tag{9.83}$$

where we note a convenient cancelation.

Now rewrite equations (9.78), (9.79), and equation (9.83) instead of (9.80),

$$\alpha' + \beta' s c_1 + \delta' s^2 c_2 + \frac{\partial \mathbf{r}}{\partial \alpha_J} \alpha'_J = 0 \qquad (9.84)$$

$$\beta' c_0 + \delta' s c_1 + \frac{\partial \mathbf{r}'}{\partial \alpha_J} \alpha'_J = Q \qquad (9.85)$$

$$-\beta' \alpha_J s c_1 + \delta' c_0 + \left(\alpha - \alpha_J \frac{\partial \mathbf{r}}{\partial \alpha_J} \right) \alpha'_J = -\mu \varepsilon'. \qquad (9.86)$$

Now solve these three equations simultaneously for α', β', and δ'.

Solve for β'

Multiplying equation (9.85) by c_0, equation (9.86) by $s c_1$, and subtracting yields

$$\beta' c_0^2 + \delta' s c_1 c_0 + \frac{\partial \mathbf{r}'}{\partial \alpha_J} \alpha'_J c_0 + \beta' \alpha_J s^2 c_1^2 - \delta' c_0 s c_1$$

$$- \left(\alpha - \alpha_J \frac{\partial \mathbf{r}}{\partial \alpha_J} \right) \alpha'_J s c_1 = Q c_0 + \mu \varepsilon' s c_1.$$

Collecting coefficients,

$$\beta' \left(c_0^2 + \alpha_J s^2 c_1^2 \right) + \delta' \left(s c_1 c_0 - c_0 s c_1 \right)$$

$$+ \alpha'_J \left(\frac{\partial \mathbf{r}'}{\partial \alpha_J} c_0 + \alpha_J s c_1 \frac{\partial \mathbf{r}}{\partial \alpha_J} - \alpha s c_1 \right) = Q c_0 + \mu \varepsilon' s c_1.$$

Note that the coefficient of β' is unity (eq. E.10) and the coefficient of δ' is zero, so we solve for β' to get

$$\beta' = Q c_0 + \mu \varepsilon' s c_1 - \alpha'_J \overbrace{\left(\frac{\partial \mathbf{r}'}{\partial \alpha_J} c_0 + \alpha_J s c_1 \frac{\partial \mathbf{r}}{\partial \alpha_J} - \alpha s c_1 \right)}^{de1}. \qquad (9.87)$$

We label the coefficient of α'_J as $de1$ for reference. We will simplify the coefficient shortly.

Solve for δ'

Multiplying equation (9.85) by $\alpha_J s c_1$, equation (9.86) by c_0, and adding yields

$$\beta' \alpha_J s c_1 c_0 + \delta' \alpha_J s^2 c_1^2 + \frac{\partial \mathbf{r}'}{\partial \alpha_J} \alpha'_J \alpha_J s c_1 - \beta' \alpha_J s c_0 c_1 + \delta' c_0^2$$

$$+ \left(\alpha - \alpha_J \frac{\partial \mathbf{r}}{\partial \alpha_J} \right) \alpha'_J c_0 = Q \alpha_J s c_1 - \mu \varepsilon' c_0.$$

Observing that the factors that are coefficients of β' cancel and the factors that are coefficients of δ' become unity, we solve for δ' to get

$$\delta' = Q\alpha_{,s}c_1 - \mu\varepsilon'c_0 - \alpha'_{,} \underbrace{\left(\frac{\partial \mathbf{r}'}{\partial \alpha_{,}}\alpha_{,s}c_1 - \frac{\partial \mathbf{r}}{\partial \alpha_{,}}\alpha_{,}c_0 + \boldsymbol{\alpha}c_0 \right)}_{de2}, \qquad (9.88)$$

where again we have introduced the notation $de2$ for the coefficient of $\alpha'_{,}$.

Solve for $\boldsymbol{\alpha}'$

Substitute equations (9.87) and (9.88) into equation (9.84) to obtain

$$\boldsymbol{\alpha}' + sc_1 \left[Qc_0 + \mu\varepsilon'sc_1 - \alpha'_{,}(de1) \right] + s^2c_2 \left[Q\alpha_{,s}c_1 \right.$$
$$\left. - \mu\varepsilon'c_0 - \alpha'_{,}(de2) \right] + \frac{\partial \mathbf{r}}{\partial \alpha_{,}}\alpha'_{,} = 0,$$

which rearranges to

$$\boldsymbol{\alpha}' + Q \left[sc_1c_0 + \alpha_{,}s^3c_1c_2 \right] + \mu\varepsilon' \left[s^2c_1^2 - s^2c_0c_2 \right]$$
$$- \alpha'_{,} \left[(de1)sc_1 + (de2)s^2c_2 - \frac{\partial \mathbf{r}}{\partial \alpha_{,}} \right] = 0.$$

We now reduce this equation term by term. For the coefficient of Q,

$$sc_1 \left(c_0 + \alpha_{,}s^2c_2 \right) = sc_1(1) = sc_1.$$

The coefficient of $\mu\varepsilon'$ is

$$s^2 \left(c_1^2 - c_0c_2 \right) = s^2c_2,$$

where we have used the elementary identities from Appendix E (eqs. E.10 and E.11),

$$c_0^2 + \alpha_{,}s^2c_1^2 = 1$$
$$c_0 + \alpha_{,}s^2c_2 = 1$$

to obtain

$$c_1^2 - c_0c_2 = c_2.$$

Substitute these results into the equation of $\boldsymbol{\alpha}'$ to obtain

$$\boldsymbol{\alpha}' + Qsc_1 + \mu\varepsilon's^2c_2 - \alpha'_{,} \left[(de1)sc_1 + (de2)s^2c_2 - \frac{\partial \mathbf{r}}{\partial \alpha_{,}} \right] = 0.$$

We now reduce the coefficient of α'_J (which we refer to below as Υ).
Expanding the coefficient using the definitions for $de1$ and $de2$,

$$
\begin{aligned}
\Upsilon = sc_1 &\left(\frac{\partial \mathbf{r}'}{\partial \alpha_J} c_0 + \alpha_J sc_1 \frac{\partial \mathbf{r}}{\partial \alpha_J} - \boldsymbol{\alpha} sc_1 \right) \\
&+ s^2 c_2 \left(\frac{\partial \mathbf{r}'}{\partial \alpha_J} \alpha_J sc_1 - \frac{\partial \mathbf{r}}{\partial \alpha_J} \alpha_J c_0 + \boldsymbol{\alpha} c_0 \right) - \frac{\partial \mathbf{r}}{\partial \alpha_J}.
\end{aligned}
$$

Collecting coefficients,

$$
\begin{aligned}
\Upsilon = \boldsymbol{\alpha} \left(s^2 c_2 c_0 - s^2 c_1^2 \right) &+ \frac{\partial \mathbf{r}}{\partial \alpha_J} \left(\alpha_J s^2 c_1^2 - \alpha_J s^2 c_2 c_0 - 1 \right) \\
&+ \frac{\partial \mathbf{r}'}{\partial \alpha_J} \left(sc_1 c_0 + \alpha_J s^3 c_1 c_2 \right),
\end{aligned}
$$

which reduces to

$$
\Upsilon = -\boldsymbol{\alpha} s^2 c_2 - \frac{\partial \mathbf{r}}{\partial \alpha_J} \left(\underbrace{1 - \alpha_J s^2 c_2}_{c_0} \right) + \frac{\partial \mathbf{r}'}{\partial \alpha_J} \left(sc_1 \right).
$$

The equation for $\boldsymbol{\alpha}'$ becomes

$$
\boldsymbol{\alpha}' + Q sc_1 + \mu \boldsymbol{\varepsilon}' s^2 c_2 - \alpha'_J \left[-\boldsymbol{\alpha} s^2 c_2 - \frac{\partial \mathbf{r}}{\partial \alpha_J} c_0 + \frac{\partial \mathbf{r}'}{\partial \alpha_J} sc_1 \right] = 0
$$

or

$$
\boldsymbol{\alpha}' = -Q sc_1 - \mu \boldsymbol{\varepsilon}' s^2 c_2 - \alpha'_J \left[\underbrace{\boldsymbol{\alpha} s^2 c_2 + \frac{\partial \mathbf{r}}{\partial \alpha_J} c_0 - \frac{\partial \mathbf{r}'}{\partial \alpha_J} sc_1}_{de3} \right]. \tag{9.89}
$$

We now reduce the coefficients of α'_J in equations (9.87), (9.88), and (9.89), which we have labeled $de1$, $de2$, and $de3$, respectively. For $de1$ we will need $\frac{\partial \mathbf{r}}{\partial \alpha_J}$ and $\frac{\partial \mathbf{r}'}{\partial \alpha_J}$. From equations (9.57) and (9.58),

$$
\mathbf{r} = \boldsymbol{\alpha} + \boldsymbol{\beta} sc_1 + \boldsymbol{\delta} s^2 c_2
$$

and

$$
\mathbf{r}' = \boldsymbol{\beta} c_0 + \boldsymbol{\delta} sc_1.
$$

Therefore,

$$\frac{\partial \mathbf{r}}{\partial \alpha_J} = \beta s \frac{\partial c_1}{\partial \alpha_J} + \delta s^2 \frac{\partial c_2}{\partial \alpha_J} \tag{9.90}$$

and

$$\frac{\partial \mathbf{r}'}{\partial \alpha_J} = \beta \frac{\partial c_0}{\partial \alpha_J} + \delta s \frac{\partial c_1}{\partial \alpha_J}. \tag{9.91}$$

Substitute equations (9.90) and (9.91) into the equation for $de1$ and collect coefficients of α, β and δ to obtain

$$de1 = \beta \left(c_0 \frac{\partial c_0}{\partial \alpha_J} + \alpha_J s^2 c_1 \frac{\partial c_1}{\partial \alpha_J} \right) + \delta \left(s c_0 \frac{\partial c_1}{\partial \alpha_J} + \alpha_J s^3 c_1 \frac{\partial c_2}{\partial \alpha_J} \right) - \alpha s c_1.$$

Previously we developed derivatives of the Stumpff functions with respect to s. Now we need partial derivatives with respect to α_J. For $n = 0$ we had (from Appendix E) the general relation

$$\frac{dc_0(z)}{dz} = -\frac{1}{2} c_1(z),$$

with $z = \alpha_J s^2$. Therefore,

$$\frac{dc_0 \left(\alpha_J s^2 \right)}{dz} \frac{dz}{d\alpha_J} = -\frac{1}{2} c_1 \left(\alpha_J s^2 \right) (s^2)$$

$$\frac{\partial c_0}{\partial \alpha_J} = -\frac{1}{2} s^2 c_1.$$

Also from Appendix E we had

$$2z \frac{dc_n(z)}{dz} = c_{n-1}(z) - n c_n(z)$$

for $n \geq 1$. With $z = \alpha_J s^2$, this gives

$$\frac{\partial c_n}{\partial \alpha_J} = \frac{1}{2\alpha_J} \left(c_{n-1} - n c_n \right).$$

Substitute these partials into $de1$ to get

$$de1 = \beta \left[c_0 \left(-\frac{1}{2} s^2 c_1 \right) + \alpha_J s^2 c_1 \left(\frac{c_0 - c_1}{2\alpha_J} \right) \right]$$
$$+ \delta \left[s c_0 \left(\frac{c_0 - c_1}{2\alpha_J} \right) + \alpha_J s^3 c_1 \left(\frac{c_1 - 2c_2}{2\alpha_J} \right) \right] - \alpha s c_1,$$

and using the elementary identity, $c_0^2 + \alpha_J s^2 c_1^2 = 1$, this becomes

$$de1 = -\frac{1}{2}\beta s^2 c_1^2 + \delta \frac{s}{2\alpha_J}\left[1 - c_0 c_1 - 2\alpha_J s^2 c_1 c_2\right] - \alpha s c_1.$$

We can simplify this equation further by using relationships for the Stumpff functions from Appendix E, equations (E.14), (E.1), and (E.13):

$$c_1(4z) = c_0(z)c_1(z)$$

$$1 = c_1(4z) + 4z c_3(4z)$$

$$c_1^2(z) = 2c_2(4z),$$

with $z = \alpha_J s^2$. We denote $c_n(4z)$ with the notation $\tilde{c}_n$. Using this notation we have

$$\tilde{c}_1 = c_0 c_1$$

$$1 = \tilde{c}_1 + 4\alpha_J s^2 \tilde{c}_3$$

$$c_1^2 = 2\tilde{c}_2,$$

and $de1$ becomes

$$de1 = -\alpha s c_1 - \beta s^2 \tilde{c}_2 + \delta s^3 \left(2\tilde{c}_3 - c_1 c_2\right).$$

Substitute this equation into equation (9.87) to finally obtain

$$\beta' = Qc_0 + \mu\varepsilon' s c_1 + \alpha'_J\left[\alpha s c_1 + \beta s^2 \tilde{c}_2 - \delta s^3 \left(2\tilde{c}_3 - c_1 c_2\right)\right]. \tag{9.92}$$

Equation (9.88) similarly reduces to

$$\delta' = Q\alpha_J s c_1 - \mu\varepsilon' c_0 + \alpha'_J\left[-\alpha c_0 + 2\beta\alpha_J s^3 \tilde{c}_3 + \frac{1}{2}\delta\alpha_J s^4 c_2^2\right] \tag{9.93}$$

and equation (9.89) reduces to

$$\alpha' = -Qs c_1 - \mu\varepsilon' s^2 c_2 - \alpha'_J\left[\alpha s^2 c_2 + 2\beta s^3 \tilde{c}_3 + \frac{1}{2}\delta s^4 c_2^2\right]. \tag{9.94}$$

9.5.2 Equations for the Temporal Elements

The solution for the temporal system is

$$y_1 = r = a + bs c_1 + \gamma s^2 c_2 \tag{9.95}$$

$$y_2 = r' = bc_0 + \gamma s c_1 \tag{9.96}$$

$$y_3 = \mu - \alpha_J r = \mu - y_5 y_1 \tag{9.97}$$

$$y_4 = t = \tau + as + bs^2 c_2 + \gamma s^3 c_3 \tag{9.98}$$

$$y_5 = \alpha_J, \tag{9.99}$$

where we introduce the notation

$$a = r_0 = y_1(0)$$

$$b = r_0' = y_2(0)$$

$$\gamma = \mu - \alpha_J a = y_3(0)$$

$$\tau = t_0 = y_4(0).$$

Note that we must again include α_J since it appears in $c_k = c_k(\alpha_J, s^2)$.

Now, from the differential equation (9.29) for the perturbed temporal system we derive a system of first-order differential equations of the form

$$\mathbf{y}' = \mathbf{f} + \mathbf{g},$$

where $\mathbf{f}$ is the unperturbed part and $\mathbf{g}$ is the perturbation. From equations (9.95)–(9.99) we get

$$y_1' = r' = y_2$$

$$y_2' = r'' = -\alpha_J r + \mu + r\mathbf{r} \cdot \mathbf{F} + 2(\sigma - V)r$$

$$y_3' = -\alpha_J r' - \alpha_J' r = -y_5 y_2 - \alpha_J' r$$

$$y_4' = t' = r = y_1$$

$$y_5' = \alpha_J'.$$

As we did for the spatial differential equations, we put the above equations in the required form $y' = f + g$,

$$y_1' = y_2 + g_1 \tag{9.100}$$

$$y_2' = y_3 + g_2 \tag{9.101}$$

$$y_3' = -y_5 \, y_2 + g_3 \tag{9.102}$$

$$y_4' = y_1 + g_4 \tag{9.103}$$

$$y_5' = g_5, \tag{9.104}$$

where we have defined the perturbations

$$g_1 = 0 \tag{9.105}$$

$$g_2 = r\mathbf{r} \cdot \mathbf{F} + 2(\sigma - V)r = \frac{1}{r}\mathbf{Q} \cdot \mathbf{r} \tag{9.106}$$

$$g_3 = -\alpha_j' r \tag{9.107}$$

$$g_4 = 0 \tag{9.108}$$

$$g_5 = \alpha_j'. \tag{9.109}$$

The differential equations for the new independent variables (the old constants $a, b, \gamma, \tau, \alpha_j$) are found from Lagrange's method to be

$$\frac{\partial \mathbf{y}}{\partial \mathbf{K}} \mathbf{K}' = \mathbf{g},$$

where

$$\mathbf{K}^T = (a, b, \gamma, \tau, \alpha_j).$$

The matrix of partial derivatives is obtained from equations (9.95)–(9.99), where $a, b, \gamma, \tau, \alpha_j$ are now functions of s.

For y_1 we have

$$\frac{\partial y_1}{\partial \mathbf{K}} = \left[\begin{array}{ccccc} \dfrac{\partial r}{\partial a} & \dfrac{\partial r}{\partial b} & \dfrac{\partial r}{\partial \gamma} & \dfrac{\partial r}{\partial \tau} & \dfrac{\partial r}{\partial \alpha_j} \end{array} \right].$$

Taking the partials we get

$$\frac{\partial y_1}{\partial \mathbf{K}} = \left[\begin{array}{ccccc} 1 & sc_1 & s^2 c_2 & 0 & \dfrac{\partial r}{\partial \alpha_j} \end{array} \right]. \tag{9.110}$$

For y_2 we have

$$\frac{\partial y_2}{\partial \mathbf{K}} = \left[\begin{array}{ccccc} \dfrac{\partial r'}{\partial a} & \dfrac{\partial r'}{\partial b} & \dfrac{\partial r'}{\partial \gamma} & \dfrac{\partial r'}{\partial \tau} & \dfrac{\partial r'}{\partial \alpha_j} \end{array} \right].$$

Taking the partials we get

$$\frac{\partial y_2}{\partial \mathbf{K}} = \left[\begin{array}{ccccc} 0 & c_0 & sc_1 & 0 & \dfrac{\partial r'}{\partial \alpha_j} \end{array} \right]. \tag{9.111}$$

For y_3 we have

$$y_3 = \mu - \alpha_j r,$$

but since

$$\gamma = \mu - \alpha_J a,$$

we can express y_3 as

$$y_3 = \gamma + \alpha_J a - \alpha_J r,$$

so the alternate expression for y_3 is

$$y_3 = \gamma + \alpha_J (a - r)$$

Taking the partials

$$\frac{\partial y_3}{\partial \mathbf{K}} = \left[\begin{array}{ccccc} \dfrac{\partial y_3}{\partial a} & \dfrac{\partial y_3}{\partial b} & \dfrac{\partial y_3}{\partial \gamma} & \dfrac{\partial y_3}{\partial \tau} & \dfrac{\partial y_3}{\partial \alpha_J} \end{array} \right],$$

but

$$\frac{\partial y_3}{\partial a} = \alpha_J \left(1 - \frac{\partial r}{\partial a} \right) = 0$$

$$\frac{\partial y_3}{\partial b} = -\alpha_J \frac{\partial r}{\partial b} = -\alpha_J s c_1$$

$$\frac{\partial y_3}{\partial \gamma} = 1 - \alpha_J \frac{\partial r}{\partial \gamma} = 1 - \alpha_J s^2 c_2 = c_0$$

$$\frac{\partial y_3}{\partial \tau} = 0.$$

So the partials are

$$\frac{\partial y_3}{\partial \mathbf{K}} = \left[\begin{array}{ccccc} 0 & -\alpha_J s c_1 & c_0 & 0 & \dfrac{\partial y_3}{\partial \alpha_J} \end{array} \right]. \tag{9.112}$$

For y_4 we have

$$\frac{\partial y_4}{\partial \mathbf{K}} = \left[\begin{array}{ccccc} \dfrac{\partial t}{\partial a} & \dfrac{\partial t}{\partial b} & \dfrac{\partial t}{\partial \gamma} & \dfrac{\partial t}{\partial \tau} & \dfrac{\partial t}{\partial \alpha_J} \end{array} \right],$$

which becomes

$$\frac{\partial y_4}{\partial \mathbf{K}} = \left[\begin{array}{ccccc} s & s^2 c_2 & s^3 c_3 & 1 & \dfrac{\partial t}{\partial \alpha_J} \end{array} \right]. \tag{9.113}$$

Since $y_5 = \alpha_J$, the last row becomes

$$\frac{\partial y_5}{\partial \mathbf{K}} = [\, 0 \quad 0 \quad 0 \quad 0 \quad 1 \,]. \tag{9.114}$$

Combining equations (9.110)–(9.114) yields

$$
\left[\frac{\partial \mathbf{y}}{\partial \mathbf{K}}\right] \mathbf{K'} = \mathbf{g} =
\begin{pmatrix}
1 & sc_1 & s^2c_2 & 0 & \frac{\partial r}{\partial \alpha_J} \\
0 & c_0 & sc_1 & 0 & \frac{\partial r'}{\partial \alpha_J} \\
0 & -\alpha_J sc_1 & c_0 & 0 & \frac{\partial y_3}{\partial \alpha_J} \\
s & s^2c_2 & s^3c_3 & 1 & \frac{\partial t}{\partial \alpha_J} \\
0 & 0 & 0 & 0 & 1
\end{pmatrix}
\begin{pmatrix}
a' \\
b' \\
\gamma' \\
\tau' \\
\alpha'_J
\end{pmatrix}
$$

$$
= \begin{pmatrix}
g_1 \\
g_2 \\
g_3 \\
g_4 \\
g_5
\end{pmatrix}.
\tag{9.115}
$$

When equation (9.115) is expanded, we obtain

$$
a' + b'sc_1 + \gamma's^2c_2 + \alpha'_J \frac{\partial r}{\partial \alpha_J} = g_1 = 0
\tag{9.116}
$$

$$
b'c_0 + \gamma'sc_1 + \alpha'_J \frac{\partial r'}{\partial \alpha_J} = g_2 = \frac{1}{r}\mathbf{Q}\cdot\mathbf{r}
\tag{9.117}
$$

$$
-b'\alpha_J sc_1 + \gamma'c_0 + \alpha'_J \frac{\partial y_3}{\partial \alpha_J} = g_3 = -\alpha'_J r
\tag{9.118}
$$

$$
a's + b's^2c_2 + \gamma's^3c_3 + \tau' + \alpha'_J \frac{\partial t}{\partial \alpha_J} = g_4 = 0
\tag{9.119}
$$

$$
\alpha'_J = g_5.
\tag{9.120}
$$

We now solve equations (9.116)–(9.118) for a', b', and γ'. Comparing these equations with equations (9.78)–(9.82) for the *spatial* system, we note that equations (9.116)–(9.118) are *identical in form* to equations (9.78)–(9.80). We can use this fact and write the differential equations for a, b, and γ. So we have

$$
a' = -\frac{1}{r}\mathbf{Q}\cdot\mathbf{r}sc_1 - \alpha'_J\left[as^2c_2 + 2bs^3\tilde{c}_3 + \frac{1}{2}\gamma s^4c_2^2\right]
\tag{9.121}
$$

$$
b' = \frac{1}{r}\mathbf{Q}\cdot\mathbf{r}c_0 + \alpha'_J\left[asc_1 + bs^2\tilde{c}_2 - \gamma s^3\left(2\tilde{c}_3 - c_1c_2\right)\right]
\tag{9.122}
$$

$$
\gamma' = \frac{1}{r}\mathbf{Q}\cdot\mathbf{r}\alpha_J sc_1 + \alpha'_J\left[-ac_0 + 2b\alpha_J s^3\tilde{c}_3 + \frac{1}{2}\gamma\alpha_J s^4c_2^2\right].
\tag{9.123}
$$

The differential equation for τ is found by substituting the derivatives given in equations (9.121)–(9.123) into equation (9.119). Also, from equations (9.95)–

(9.99) we obtain

$$\frac{\partial t}{\partial \alpha_J} = bs^2 \frac{\partial c_2}{\partial \alpha_J} + \gamma s^3 \frac{\partial c_3}{\partial \alpha_J}. \tag{9.124}$$

Omitting the algebra (which is similar to that in obtaining the spatial elements) we finally obtain

$$\tau' = \frac{1}{r}\mathbf{Q} \cdot \mathbf{r}s^2 c_2 + \alpha'_J \left[as^3 c_3 + \frac{1}{2}bs^4 c_2^2 - 2\gamma s^5(c_5 - 4\tilde{c}_5) \right]. \tag{9.125}$$

9.6 Summary of the Equations

The complete system of differential equations is as follows:

$$\boldsymbol{\alpha'} = -Qsc_1 - \mu\varepsilon's^2c_2 - \alpha'_J \left[\boldsymbol{\alpha}s^2c_2 + 2\boldsymbol{\beta}s^3\tilde{c}_3 + \frac{1}{2}\boldsymbol{\delta}s^4c_2^2 \right]$$

$$\boldsymbol{\beta'} = Qc_0 + \mu\varepsilon'sc_1 + \alpha'_J \left[\boldsymbol{\alpha}sc_1 + \boldsymbol{\beta}s^2\tilde{c}_2 - \boldsymbol{\delta}s^3 \left(2\tilde{c}_3 - c_1c_2\right) \right]$$

$$\boldsymbol{\delta'} = Q\alpha_J sc_1 - \mu\varepsilon'c_0 + \alpha'_J \left[-\boldsymbol{\alpha}c_0 + 2\boldsymbol{\beta}\alpha_J s^3\tilde{c}_3 + \frac{1}{2}\boldsymbol{\delta}\alpha_J s^4c_2^2 \right]$$

$$a' = -\frac{1}{r}\mathbf{Q} \cdot \mathbf{r}sc_1 - \alpha'_J \left[as^2c_2 + 2bs^3\tilde{c}_3 + \frac{1}{2}\gamma s^4c_2^2 \right]$$

$$b' = \frac{1}{r}\mathbf{Q} \cdot \mathbf{r}c_0 + \alpha'_J \left[asc_1 + bs^2\tilde{c}_2 - \gamma s^3 \left(2\tilde{c}_3 - c_1c_2\right) \right]$$

$$\gamma' = \frac{1}{r}\mathbf{Q} \cdot \mathbf{r}\alpha_J sc_1 + \alpha'_J \left[-ac_0 + 2b\alpha_J s^3\tilde{c}_3 + \frac{1}{2}\gamma\alpha_J s^4c_2^2 \right]$$

$$\tau' = \frac{1}{r}\mathbf{Q} \cdot \mathbf{r}s^2c_2 + \alpha'_J \left[as^3c_3 + \frac{1}{2}bs^4c_2^2 - 2\gamma s^5(c_5 - 4\tilde{c}_5) \right]$$

$$\alpha'_J = 2(-\mathbf{r}' + r\,\boldsymbol{\omega} \times \mathbf{r}) \cdot \mathbf{P}$$

$$\sigma' = r\boldsymbol{\omega} \cdot \mathbf{r} \times \mathbf{F},$$

where

$$\mu\varepsilon' = 2(\mathbf{r}' \cdot \mathbf{F})\mathbf{r} - (\mathbf{r} \cdot \mathbf{F})\mathbf{r}' - (\mathbf{r}' \cdot \mathbf{r})\mathbf{F}$$

$$\mathbf{Q} = r^2\mathbf{F} + 2(\sigma - V)\mathbf{r},$$

which is a system of fifteen differential equations. We also observe from equation (9.45) that

$$\gamma + a\,\alpha_J = \mu = constant,$$

which amounts to an integral of the motion; that is, the element γ can always be computed from the elements a and α_J. This in effect reduces the number of differential equations from fifteen to fourteen.

When the perturbations are zero (that is, $\mathbf{F} = 0, \mathbf{Q} = 0$, and $\mathbf{P} = 0$), the system of differential equations reduces to

$$\mathbf{z}' = 0,$$

where we use $\mathbf{z}$ as a vector representing the entire array of dependent variables,

$$\mathbf{z}^T = \left(\boldsymbol{\alpha}^T, \boldsymbol{\beta}^T, \boldsymbol{\delta}^T, a, b, \gamma, \tau, \alpha_J, \sigma \right).$$

Note that in contrast to the Cowell and Encke methods discussed in §9.1, this system of differential equations for the orbital elements has no propagation error growth when the perturbations are zero.

9.7 Sperling-Burdet Method—Examples

In this section some numerical examples are presented. These examples are all solutions of the system of differential equations, which are summarized in §9.6.

The first example was presented by Stiefel and Scheifele [64] and involves a geocentric satellite that is perturbed by an oblate Earth and the Moon. Here the oblateness (or second zonal harmonic, J_2) is modeled by a potential that is a function of position only. The third-body (or lunar) perturbation is an explicit function of time that will be included as a $\mathbf{P}$-type acceleration as defined in §9.3. The solution by the Sperling-Burdet method will be directly compared to other published solutions.

The second example is that of a satellite at a stable libration point in an idealized Earth-Moon system. This example is of course normally associated with the restricted three-body problem [69], but here we transform the problem into an inertial system and treat the motion of the satellite as a perturbed two-body problem. The perturbation by the Moon is treated as that due to a potential that is an explicit function of time (refer to §8.7.3 and §8.7.5). The beauty of this example is that the numerical solution can be compared directly to a well-known analytical solution. Burdet used this example in his original work [25].

The third example is that of a satellite perturbed by a continuous radial thrust. This problem was shown to have an analytical solution by Tsien in the 1950s (see Battin [5] for an exquisite reproduction of Tsien's solution). In the

Table 9.2   Comparison of several methods for the oblate Earth plus Moon problem.

Method	Ref. [64]	Sperling-Burdet [17]	Kustaanheimo-Stiefel [14]	Cowell [19]
x_1 (km)	−24219.050	−24218.818	−24219.002	−24182.152
x_2 (km)	227962.106	227961.915	227962.429	227943.989
x_3 (km)	129753.442	129753.343	129753.822	129744.270
Steps/Rev	500	62	62	240
RSS Error		0.318	0.501	42.5

Sperling-Burdet solution, the radial thrust is modeled as a **P**-type acceleration. Here, as in the second example, we have the opportunity to compare directly to an analytical solution.

9.7.1 Oblate Earth Plus the Moon

The first example is that of a highly eccentric ($e = 0.95$) orbit about the Earth. The orbit is subject to the J_2 (Earth oblateness) perturbing potential, which is conservative, plus lunar perturbations. This example was also computed by Stiefel and Scheifele [64] with extremely high precision and will be used as a reference. The numerical method used in all cases was the $RKF4(5)$, which we cover in chapter 10.

The initial position and velocity for the satellite are

$$\mathbf{r}_0 = (-5888.9727 \,\text{km})\,\hat{\mathbf{j}} - (3400.0 \,\text{km})\,\hat{\mathbf{k}}$$
$$\dot{\mathbf{r}}_0 = (10.691338 \,\text{km/sec})\,\hat{\mathbf{i}},$$

where the coordinate system axes x_1 and x_2 (and corresponding unit vectors $\hat{\mathbf{i}}$ and $\hat{\mathbf{j}}$) are fixed in the Earth equatorial plane and the x_3 (and $\hat{\mathbf{k}}$) axis is perpendicular to the equatorial plane. The final satellite positions are after approximately fifty revolutions (288.12768941 days).

Three solutions are shown in table 9.2, each compared to the reference by means of the root sum square of the difference from the reference solution; that is,

$$\text{RSS} = \sqrt{\|\mathbf{r}_{ref} - \mathbf{r}\|}.$$

The Earth oblateness and lunar models used are somewhat idealized and are taken from Stiefel and Scheifele [64]. The Earth oblateness perturbations were computed from the potential

$$V = \frac{3}{2}(\text{GE}) \, J_2 \frac{a_e^2}{r^3} \left(\frac{x_3^2}{r^2} - \frac{1}{3} \right),$$

which was introduced in §8.7.2 and will also be discussed in chapter 11. Note that

$\mu = GE = 398601 \text{ km}^3/\text{sec}^2$ (gravitational constant of the Earth)

$a_e = 6371.22 \text{ km}$ (equatorial radius of the Earth)

$J_2 = 1.08265 \times 10^{-3}$ (second harmonic of the geopotential).

The lunar perturbation (see §8.7.5 and chapter 11) was computed from

$$\mathbf{P} = -G \, M \left[\frac{\mathbf{r} - \rho}{\|\mathbf{r} - \rho\|^3} + \frac{\rho}{\rho^3} \right],$$

and the idealized lunar ephemeris is given by

$$\rho = \rho \left(\sin \Omega t \, \hat{\mathbf{i}} - \frac{\sqrt{3}}{2} \cos \Omega t \, \hat{\mathbf{j}} - \frac{1}{2} \cos \Omega t \, \hat{\mathbf{k}} \right),$$

where

$\rho = 384400 \text{ km}$ (the Earth-Moon distance)

$\Omega = 2.665315780887 \times 10^{-6} \text{ rad/sec}$ (Moon orbital rate)

$GM = 4902.66 \text{ km}^3/\text{sec}^2$ (gravitational constant of the Moon).

Note that the Sperling-Burdet method (Bond and Fraietta [17]) and the Kustaanheimo-Stiefel method (Bond [14]) give nearly the same solutions as the reference (RSS = 0.318 km and 0.501 km, respectively) and require only an average of 62 numerical steps per revolution. The Cowell method (Bond and Hanssen [19]), which is the direct integration of equation (9.11), has an RSS of 42.5 km and requires 240 numerical steps per revolution.

9.7.2 Stable Libration Points

The example is that of a mass particle at a stable libration point in the Earth-Moon system. If the Earth and Moon are in circular orbits about their center of mass, then, from the theory of the restricted three-body problem (Szebehely [69]), if the infinitesimal mass is placed 60° ahead of, or behind, the Moon in the Earth-Moon plane and at the same distance (from Earth) and circular speed of the Moon, then the infinitesimal mass will remain at either of these points indefinitely.

The initial conditions for this problem are transformed to an inertial system, and the motion of the infinitesimal mass will be treated as a perturbed two-body problem. The x_1-axis of an inertial system is the line from the Earth to Moon at time $t = 0$. The x_3-axis of this inertial system is normal to the Earth-Moon plane. The ephemeris of the Moon is thus given by

$$\boldsymbol{\rho} = \rho(\cos \omega t \,\hat{\mathbf{i}} + \sin \omega t \,\hat{\mathbf{j}}).$$

We also choose for the Earth and Moon gravitational constants,

$$\mu = GE = 398601 \,\text{km}^3/\text{sec}^2$$

$$GM = 4902.66 \,\text{km}^3/\text{sec}^2.$$

Note that ω is the mean motion of the Earth-Moon system and is found from

$$\omega^2 = \frac{GE + GM}{\rho^3}.$$

The vector $\boldsymbol{\omega}$ is normal to the Earth-Moon plane, therefore

$$\boldsymbol{\omega} = \omega \hat{\mathbf{k}}.$$

The third-body gravitational potential can be expressed as

$$V = -GM \left(\frac{1}{\Delta} - \frac{\mathbf{r} \cdot \boldsymbol{\rho}}{\rho^3} \right),$$

which is in the conventional notation (see the previous example). The gradient of the third-body gravitational potential is

$$\frac{\partial V}{\partial \mathbf{r}} = GM \left(\frac{\mathbf{r} - \boldsymbol{\rho}}{\Delta^3} + \frac{\boldsymbol{\rho}}{\rho^3} \right),$$

where

$$\Delta = \|\mathbf{r} - \boldsymbol{\rho}\|.$$

At time $t = 0$ we choose the initial conditions of a satellite in this inertial system to be

$$\mathbf{r}_0 = \frac{1}{2}\rho \left(\hat{\mathbf{i}} + \sqrt{3}\,\hat{\mathbf{j}} \right)$$

Table 9.3 Comparison of the Sperling-Burdet numerical solution to analytic solution at a stable libration point in the Earth-Moon system.

t (lunar period)	x_1 (km)	Δx_1 (m)	x_2 (km)	Δx_2 (m)
0	192200.0000	0	332900.1652	0
250	192200.0022	2.2	332900.1650	−0.2
500	192200.0020	2.0	332900.1660	0.8
750	192200.0056	5.6	332900.1662	1.0
1000	192200.0033	3.3	332900.1681	2.9

and

$$\dot{r}_0 = \frac{1}{2}\omega\rho\left(\hat{j} - \sqrt{3}\,\hat{i}\right).$$

This initial position corresponds to the position of a stable Lagrangian point in the idealized Earth-Moon system. Since the Lagrangian point is at the same distance and moves with the same rate as the Moon, after each lunar period,

$$T = \frac{2\pi}{\omega} = 27.28459145 \text{ days},$$

the position of the satellite should once again be at the initial conditions in the inertial system. The deviations (Δx_1, Δx_2) of the position of the satellite from the Lagrangian point is thus a direct measure of the numerical error in the performance of the Sperling-Burdet method. Further, since the Lagrangian libration point is stable, there is no propagation error. Thus the error is purely due to the truncation error incurred by the integration method employed. This problem was also solved by the Sperling-Burdet method [10] and compared to the analytic solution in table 9.3. The numerical integration method used was a fixed-step Runge-Kutta 4th order using 80 steps per lunar period.

9.7.3 Continuous Radial Thrust

The third example is Tsien's solution to a radial thrust problem, recently described by Battin [5] and also applied by Murad [56]. A spacecraft initially in a circular orbit has a constant, radial, sufficiently large acceleration applied until it attains a parabolic (that is, escape) velocity. The distance and time at escape can be computed analytically. The authors gratefully acknowledge the assistance of Dr. R. G. Gottlieb [35] for computing the analytical solution. The perturbation has the form

$$\mathbf{P} = A\frac{\mathbf{r}}{r},$$

Table 9.4   Sperling-Burdet numerical solution compared with Tsien's analytical solution.

Method	Time (sec)	Eccentricity	Position Mag. (km)
Analytical	$t = 0$	$e = 0$	$r = 6800$
Sperling-Burdet	$t = 0$	$e = 0$	$r = 6800$
Analytical	$t = 12000$	$e = 1$	$r = 30682.72380$
Sperling-Burdet	$t = 12000$	$e = 1$	$r = 30682.72393$

where A is the specified magnitude of **P**. The initial conditions are specified to those of a spacecraft in an exact circular orbit of radius r_0. In the geocentric inertial coordinate system, we choose the initial position and velocity to be

$$\mathbf{r}_0 = r_0 \,\hat{\mathbf{i}}$$

$$\dot{\mathbf{r}}_0 = \sqrt{\frac{\mu}{r_0}}\,\hat{\mathbf{j}},$$

where $\hat{\mathbf{i}}$ and $\hat{\mathbf{j}}$ are unit vectors along the positive x_1 and $\hat{\mathbf{j}}$ axes, respectively. Also,

$$\mu \,(= GE) = 398600.47 \text{ km}^3/\text{sec}^2$$

$$r_0 = 6800 \text{ km}$$

$$A = 1.22719913916381 \times 10^{-3} \text{ km/sec}^2.$$

This problem was numerically solved by the Sperling-Burdet method [10] to the specified time of $t = 12000.00$ sec. The final conditions are given in table 9.4. Note that with an average time-step size of approximately 105 sec per step, the error in position magnitude was 0.13 meters. The numerical integration method used was a fixed-step Runge-Kutta 4th order algorithm run for 114 steps.

10

RUNGE-KUTTA METHODS

10.1 Introduction

The differential equations of motion of the perturbed two-body problem are in general unsolvable in the sense of the definition of a solvable system in chapter 2. The inclusion of perturbations has rendered the system unsolvable. As discussed in chapter 8, we must numerically solve the perturbed system

$$\ddot{\mathbf{r}} + \frac{\mu}{r^3}\mathbf{r} = \mathbf{F}$$

either directly in cartesian coordinates or in a transformed version that led to differential equations for orbital elements such as those developed in the previous chapter.

A number of approaches can be used for solving systems of ordinary differential equations (ODEs), which includes the classes of differential equations that are of interest to us. Ordinary differential equations are of the form

$$\frac{d\mathbf{x}}{dt} = \mathbf{f}(\mathbf{x}, t); \quad \mathbf{x}_0 = \mathbf{x}(t_0),$$

where $\mathbf{x}$ represents the array of dependent variables and t is the independent variable.

The independent variable is most often time but as we have seen the independent variable can be fictitious time s, which we introduced in chapter 9. Also, the dependent variables can be cartesian coordinates and velocities or orbital elements.

There are several suitable methods for solving ordinary differential equations. These methods always involve evaluations of the $\mathbf{f}(\mathbf{x}, t)$ terms of the ODEs. One class of methods, called the *multistep methods*, relies on the evaluation of $\mathbf{f}(\mathbf{x}, t)$ over several increments (called *steps*) of the independent variable. The solution at the end of any step is computed from several preceding values of the solution

[60]. The starting set of these values must be computed by another method, quite often one of the Runge-Kutta methods discussed below.

Multistep methods yield very precise solutions, but their major disadvantage is the computer overhead in storing the values of the $f(x, t)$ over several steps. Another serious disadvantage is that large changes in perturbations cause disruptions in the sequence of evaluations of $f(x, t)$ and the solutions must then be restarted. For example, the application of a maneuvering thruster cannot be accommodated unless the solution is reinitialized.

Another class of numerical methods, and the one which we will emphasize here, are called "single-step methods." Single-step methods require only that the initial conditions $x(t_0)$ be known and that the step size be known. The $f(x, t)$ terms are then evaluated in such a sequence that they precisely duplicate the solutions of a Taylor series solution over the same step. The single-step methods are used extensively in situations where perturbations can rapidly become large (atmospheric entry, for example) and where thrusters are likely to be applied.

The single-step methods usually have the hyphenated name of Runge-Kutta methods. Low-order single-step methods were first developed by Runge in 1895 and Kutta in 1901. The advent of the digital computer inspired the development of higher-order Runge-Kutta type methods as well as Fehlberg [32] methods, which also permit the computation of the step size for the next step. More recently, Bettis [6] developed the embedded Runge-Kutta methods that have the feature of variable order as well as variable step; and Horn [39] developed Runge-Kutta methods that compute the numerical solution within the step. In what follows we shall concentrate on the Runge, Kutta, and Fehlberg methods that usually are referred to by the acronym *RK* (for Runge-Kutta) or *RKF* (Runge-Kutta-Fehlberg).

10.2 Error Classification

When a system of ordinary differential equations is solved by a numerical method, errors in the solution always occur, so we can never obtain the "true" solution by a numerical method. The solution we actually see is called the *numerical solution*. There are several sources (Thomas [73]) of error that contribute to the difference, which we call the *global error* (*GE*) between the true and numerical solutions. One source of error, the *round-off error*, we will dispense with immediately. Round-off errors occur because computers use arithmetic with finite precision. The solution for the minimization of round-off error is to use the maximum precision of the computer, or to use another computer that uses a greater precision to represent numbers. The problem of round-off error is really a problem of limits of technology or budgets.

We will concentrate on error sources that we can control at least to some degree. The first error source in this category is *propagation error* (*PE*), which we have already discussed in §9.1. Clearly, from §9.1 on Cowell and Encke methods and from §9.6 on the Sperling-Burdet method, *PE* is most effectively controlled by the choice of the system of differential equations. *PE* is related to the stability characteristics of the system of differential equations. It could occur even if our system of differential equations had an analytical solution. As an example, we will borrow from the famous restricted three-body problem [69]. It is well known that this problem has three analytical solutions that are unstable and two analytical solutions (under a condition on the mass ratios of two of the three bodies) that are stable. Here we are using the term "stability" in the Lyapunov sense. Lyapunov stability is very elegantly defined in Szebehely [69] as well as by Kennedy [46].

It is convenient to think of Lyapunov stability as related to propagation error using a cryptic definition from Stiefel and Scheifele [64], which we paraphrase: Lyapunov stability means that small variations in initial conditions produce only small variations in the solution of the differential equations.

The second source of error in the category of those we can control is *truncation error* (*TE*), which occurs because the solution algorithm is only an approximation to the solution of the differential equations of motion. For numerical methods of the Runge-Kutta type, we shall see that the solution algorithm is equivalent to a Taylor series solution through some specified power (which is called the *order*) of an increment, called the *step size,* of the independent variable. In general, *TE* can be reduced by increasing the order, that is, the number of terms in the solution algorithm. In §10.4 of this chapter we will see that further control over *TE* can be attained by allowing the step size to vary.

We start with a system of ordinary differential equations having the general form,

$$\frac{d\mathbf{x}}{dt} = \mathbf{f}(t, \mathbf{x}),$$

with the initial conditions

$$\mathbf{x}_0 = \mathbf{x}(t_0).$$

After the kth step (at time t_k) in a numerical algorithm to solve the system, we *define* three solutions: $\boldsymbol{\xi}(t_k)$ (the *true* solution), $\mathbf{x}(t_k)$ (the *numerical* solution), and $\mathbf{u}(t_k)$ (the *local* solution). Note that $\mathbf{u}(t_k) = \mathbf{x}(t_k)$.

Using the definition

$$t_{k+1} = t_k + h$$

$$h = \text{step size},$$

we define the following errors at t_{k+1}:

- **TE** is the local or truncation error, which is the difference between the local and numerical solutions,

$$\mathbf{TE}(t_{k+1}) = \mathbf{u}(t_{k+1}) - \mathbf{x}(t_{k+1}).$$

- **GE** is the global error, which is the difference between the true and the numerical solutions,

$$\mathbf{GE}(t_{k+1}) = \boldsymbol{\xi}(t_{k+1}) - \mathbf{x}(t_{k+1}).$$

- **PE** is the propagation error, which is the difference between the true and local solutions,

$$\mathbf{PE}(t_{k+1}) = \boldsymbol{\xi}(t_{k+1}) - \mathbf{u}(t_{k+1}).$$

From these definitions we arrive at

$$\mathbf{GE}(t_{k+1}) = \mathbf{PE}(t_{k+1}) + \mathbf{TE}(t_{k+1}).$$

This equation is more complicated than it appears. Since **PE** is the difference between the two neighboring solutions $\boldsymbol{\xi}$ and $\mathbf{u}$, it can in principle be found from the variational equation (recall the discussion in §9.1),

$$\mathbf{PE}(t_{k+1}) = \left[e^{\int_0^h A \, dt} \right] \mathbf{PE}(t_k),$$

where

$$A = \frac{\partial \mathbf{f}}{\partial \mathbf{x}}.$$

But $\mathbf{PE}(t_k)$ consists of *all* the error accumulated up to t_k. Therefore,

$$\mathbf{PE}(t_k) = \mathbf{GE}(t_k).$$

As in §9.1 for a small step h, we can consider the matrix A to be constant over the step h, and **PE** then becomes

$$\mathbf{PE}(t_{k+1}) = e^{A h} \, \mathbf{GE}(t_k).$$

If one of the eigenvalues of A is positive and real, then **PE** will always increase. For the rare "stable" case, as was one of the examples given in chapter 9, **PE** will not grow and the only error that occurs is the truncation error (**TE**).

10.3 Runge-Kutta Fixed Step

In this section and the following (§10.4), we will briefly present the approach to the development of the Runge-Kutta methods. To this end we will change notation to that of Fehlberg [32]. First, we will develop the methods for only a single differential equation, as in equation (10.1) below. Thus, for these sections we drop the vector notation completely. When we discuss the application of the methods to a system of differential equations, as we shall in §10.4.2, we will return to using vector notation. Second, we will write the differential equation with the independent variable (t) before the dependent variable (x), as shown in equation (10.1).

Define the current time step as t_0. For the solution of the ordinary differential equation,

$$\frac{dx}{dt} = f(t, x), \tag{10.1}$$

where

$$x(t_0) = x_0.$$

Following Fehlberg [32] we define the Runge-Kutta (RK) algorithm of order p and $r + 1$ stages at $t_0 + h$, where h is a fixed step in time, as follows:

$$x(t_0 + h) = x_0 + h \sum_{\kappa=0}^{r} c_\kappa f_\kappa + \mathcal{O}(h^{p+1}), \tag{10.2}$$

where $x(t_0+h)$ is the numerical solution and $x_0 + h \sum c_k f_k$ is the local solution $u(t_0 + h)$. The term $\mathcal{O}(h^{p+1})$ is the usual notation for terms of order h^{p+1} and higher.

Also,

$$f_0 = f(t_0, x_0) \tag{10.3}$$

$$f_\kappa = f\left(t_0 + \alpha_\kappa h, \ x_0 + h \sum_{\lambda=0}^{\kappa-1} \beta_{\kappa\lambda} f_\lambda\right) \tag{10.4}$$

$$\kappa = 1, 2, \ldots, r.$$

The value $r + 1$ is the number of stages that is the number of evaluations of the right side of equation (10.1). Observe that the order of the algorithm is p and is identical to that of the Taylor's series solution given by

$$x(t_0 + h) = \sum_{\nu=0}^{p} \frac{1}{\nu!} h^\nu X_\nu, \tag{10.5}$$

where

$$X_0 = x_0, \qquad X_1 = f(t_0, x_0) = \dot{x}_0,$$

and in general for $\nu = 1, 2, 3, 4, \ldots$,

$$X_\nu = \left. \frac{d^{\nu-1} f}{dt^{\nu-1}} \right|_{t_0, x_0}.$$

The coefficients c_κ, α_κ, and $\beta_{\kappa\lambda}$ are chosen such that the solution of equation (10.1) as computed by the *RK* algorithm (eqs. 10.2, 10.3, and 10.4) is *identical* to that which would be computed by the Taylor's series (eq. 10.5) through the order p.

10.3.1 Runge-Kutta First Order

Since this is a first-order solution, $p = 1$, $r + 1 = 1$, and so $r = 0$. The *RK*1 equation is, from equations (10.2), (10.3), and (10.4),

$$x(t_0 + h) = x_0 + h c_0 f_0 + \mathcal{O}(h^2),$$

where

$$f_0 = f(t_0, x_0).$$

The numerical solution is also given by the first-order Taylor's series expansion,

$$x(t_0 + h) = x_0 + h f(t_0, x_0) + \mathcal{O}(h^2).$$

Comparing coefficients, we see that $c_0 = 1$. So the *RK*1 algorithm gives the solution

$$x(t_0 + h) = x_0 + h f_0 + \mathcal{O}(h^2).$$

For a system of one dependent variable, the *RK*1 (numerical) solution is illustrated along with the local solution in figure 10.1.

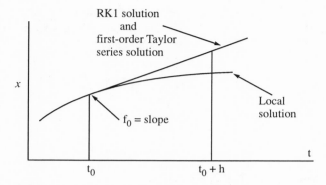

Figure 10.1 First-order Runge-Kutta solution.

10.3.2 Runge-Kutta Second Order

For RK methods above first order, we must use a Taylor series expansion in two variables. The following formula is presented for this purpose.

Given a function $f(u, v)$, we expand about the point $u_0 = a$, $v_0 = b$, and define

$$u = u_0 + k_1 = a + k_1$$
$$v = v_0 + k_2 = b + k_2.$$

With these definitions the expansion is given by

$$f(u, v) = f(a + k_1, b + k_2),$$

which becomes

$$f(u, v) = f(a, b) + \left[\left(k_1 \frac{\partial}{\partial u} + k_2 \frac{\partial}{\partial v} \right) f(u, v) \right]_{a,b} + \ldots$$
$$+ \frac{1}{n!} \left[\left(k_1 \frac{\partial}{\partial u} + k_2 \frac{\partial}{\partial v} \right)^n f(u, v) \right]_{a,b} + \ldots$$

For a second-order system we have $p = 2, r + 1 = 2$, so $r = 1$. The $RK2$ solution given by equations (10.2) through (10.4) is

$$x(t_0 + h) = x_0 + h(c_0 f_0 + c_1 f_1) + \mathcal{O}(h^3),$$

where

$$f_0 = f(t_0, x_0)$$
$$f_1 = f_\kappa|_{\kappa=1} = f(\underbrace{t_0 + \alpha_1 h,\ x_0 + h\beta_{1,0}f_0}_{\kappa=1, \lambda=0}).$$

Expanding f_1 in a double Taylor's series to order h^2, we obtain

$$f_1 = f_0 + \left[\left(\alpha_1 h \frac{\partial}{\partial t} + h\beta_{1,0}f \frac{\partial}{\partial x} \right) f(t, x) \right]_{t_0, x_0} + \mathcal{O}(h^2)$$
$$= f_0 + h \left(\alpha_1 f_t + \beta_{1,0}f f_x \right)\big|_{t_0, x_0} + \mathcal{O}(h^2).$$

Note that we only need f_1 to $\mathcal{O}(h^2)$, since f_1 will be multiplied by h in the *RK2* formula.

Now substitute f_0 and f_1 into the *RK2* solution to get

$$x(t_0 + h) = x_0 + h \left[c_0 f + c_1 f + c_1 h \left(\alpha_1 f_t + \beta_{1,0}f f_x \right) \right]_{t_0, x_0} + \mathcal{O}(h^3)$$
$$= x_0 + h \left(c_0 + c_1 \right) f_0 + h^2 \left(c_1 \alpha_1 f_t + c_1 \beta_{1,0}f f_x \right)_{t_0, x_0} + \mathcal{O}(h^3).$$

The Taylor's series solution from equation (10.5) is

$$x(t_0 + h) = x_0 + hX_1 + \frac{1}{2}h^2 X_2 + \mathcal{O}(h^3)$$
$$= x_0 + hf_0 + \frac{1}{2}h^2 \left(\underbrace{f_t + f_x f}_{df/dt} \right)_{t_0, x_0} + \mathcal{O}(h^3).$$

Comparing the *RK2* solution and the Taylor's series solution, we see that the coefficients are

$$hf_0 : c_0 + c_1 = 1$$
$$h^2 : c_1 \alpha_1 f_t + c_1 \beta_{1,0}f f_x = \frac{1}{2}(f_t + f_x f)$$
$$h^2 f_t : c_1 \alpha_1 = \frac{1}{2}$$
$$h^2 f_x f : c_1 \beta_{1,0} = \frac{1}{2}.$$

So we have a system of *three equations* of condition and *four unknowns*:

$$c_0 + c_1 = 1$$
$$c_1\alpha_1 = \frac{1}{2}$$
$$c_1\beta_{1,0} = \frac{1}{2}.$$

To determine the four coefficients $c_0, c_1, \alpha_1,$ and $\beta_{1,0}$, solve for $c_0, c_1,$ and $\beta_{1,0}$ in terms of α_1 to obtain

$$\beta_{1,0} = \alpha_1$$
$$c_1 = \frac{1}{2\alpha_1}$$
$$c_0 = 1 - c_1 = 1 - \frac{1}{2\alpha_1}.$$

α_1 is a *free* parameter. For example, if we choose $\alpha_1 = 1$, then $\beta_{1,0} = 1, c_1 = \frac{1}{2}$, and $c_0 = \frac{1}{2}$. Clearly, this solution for the unknowns $\beta_{1,0}, c_1,$ and c_0 is *not* unique. The *RK2* algorithm for $\alpha_1 = 1$ is

$$x(t_0 + h) = x_0 + \frac{1}{2}h(f_0 + f_1) + \mathcal{O}(h^3),$$

where

$$f_0 = f(t_0, x_0)$$
$$f_1 = f(t_0 + h, x_0 + hf_0).$$

Comment

- For *RK* methods the number of unknowns increases with the order more rapidly than does the number of equations of condition. Thus, as the order increases, the number of free parameters increases.

10.4 Runge-Kutta Variable Step

This approach is due to Fehlberg [32], who first developed the technique of estimating step size from truncation error. We consider two adjacent-order Runge-Kutta solutions from equations (10.2)–(10.4) for the ordinary differential equation (10.1). We call the higher-order solution $\hat{x}(t_0+h)$, and the lower-order solution $x(t_0 + h)$.

The notation for the higher-order solution is widely used (see, for example, Stoer and Bulirsch [65]). It conflicts, however, with the notation for unit vectors (e.g., $\hat{\mathbf{x}}$) when working in several dimensions, which is also a widely used notation. We therefore caution students to keep in mind that in this section references with "ˆ" are *not* unit vectors.

The *RK* algorithm, equation (10.2), gives for the two solutions,

$$x(t_0 + h) = x_0 + h \sum_{\kappa=0}^{q} c_\kappa f_\kappa + \mathcal{O}(h^{p+1}) \tag{10.6}$$

$$\hat{x}(t_0 + h) = x_0 + h \sum_{\kappa=0}^{r} \hat{c}_\kappa f_\kappa + \mathcal{O}(h^{p+2}), \tag{10.7}$$

where

$$f_0 = f(t_0, x_0) \tag{10.8}$$

$$f_\kappa = f\left(t_0 + \alpha_\kappa h, \; x_0 + h \sum_{\lambda=0}^{\kappa-1} \beta_{\kappa\lambda} f_\lambda\right) \tag{10.9}$$

$$\kappa = 1, 2, \ldots, \tau$$

$$\tau = MAX(q, r).$$

Note that the two *RK* solutions have the same α_κ and $\beta_{\kappa\lambda}$ coefficients, but different c_κ.

The difference between the two solutions, equations (10.6) and (10.7), is an *estimate of the truncation error* for the lower-order method. This should be clear from the Taylor's series solutions of order p and $p+1$, which are identical to those of the two Runge-Kutta solutions. For illustration from the Taylor's series, equation (10.5),

$$x(t_0 + h) = \sum_{\nu=0}^{p} \frac{1}{\nu!} h^\nu X_\nu, \tag{10.10}$$

but also

$$\hat{x}(t_0 + h) = \sum_{\nu=0}^{p+1} \frac{1}{\nu!} h^\nu X_\nu \tag{10.11}$$

$$= \sum_{\nu=0}^{p} \frac{1}{\nu!} h^\nu X_\nu + \frac{1}{(p+1)!} h^{p+1} X_{p+1}.$$

By subtracting equation (10.10) from (10.11), the estimate for the truncation error (*TE*) is

$$x(t_0 + h) - \hat{x}(t_0 + h) = -\frac{h^{p+1} X_{p+1}}{(p+1)!}. \qquad (10.12)$$

The actual estimate of the truncation error is based on the *RK* solutions. We subtract equation (10.7) from equation (10.6) to get

$$TE = x(t_0 + h) - \hat{x}(t_0 + h) = h \sum_{\kappa=0}^{\tau} (c_\kappa - \hat{c}_\kappa) f_\kappa + \mathcal{O}(h^{p+2}). \qquad (10.13)$$

From equation (10.13) we see that during a single fixed step h the truncation error is proportional to h^{p+1}. That is,

$$TE = \eta h^{p+1}, \qquad (10.14)$$

where η is a constant of proportionality. Now over the next step (h_1), we assume that η does not change, and further we specify the maximum truncation error allowed to be *TOL*. Then, based upon equation (10.14),

$$TOL = \eta h_1^{p+1}. \qquad (10.15)$$

Now take the ratio of equation (10.15) to equation (10.14) and solve for the next step (h_1),

$$h_1 = h \left(\frac{TOL}{TE} \right)^{\frac{1}{p+1}}, \qquad (10.16)$$

with the requirement that $TE \leq TOL$.

In view of Fehlberg's significant contribution to the Runge-Kutta methods, the Runge-Kutta variable step methods are sometimes called $RKFp(p+1)$, where p is the order of the method.

10.4.1 The RKF1(2)

We can use the results of the examples for *RK*1 and *RK*2, where we substitute the $\hat{x}$ notation into the *RK*2 solution. Subtract the *RK*2 solution from the *RK*1 solution to get the truncation error,

$$x(t_0 + h) - \hat{x}(t_0 + h) = x_0 + h f_0 - \left[x_0 + \frac{1}{2} h (f_0 + f_1) \right]$$

$$= \frac{1}{2} h (f_0 - f_1).$$

From equation (10.14),

$$TE = \mathcal{O}(h^2) = \eta h^2.$$

Now we make the following assumptions:

1. The proportionality constant η is also constant over the next step h_1.
2. A maximum error *TOL* will be allowed during the next step h_1.

We have therefore

$$TOL = \eta h_1^2,$$

and from the ratio

$$\frac{TE}{TOL} = \frac{\eta h^2}{\eta h_1^2},$$

we can solve for h_1,

$$h_1 = h\sqrt{\frac{TOL}{TE}}.$$

Since the development of any $RKFp(p+1)$ method is lengthy when $p > 1$, we will illustrate the development with the $RKF4(5)$ [32] in its final form.

10.4.2 Algorithm No. 8: RKF4(5)

Given The n-dimension vectors $\mathbf{x}, \mathbf{f}, \Delta\mathbf{x}; t_0, \mathbf{x}_0, h$, and *TOL*.
Find h_1 and $\mathbf{x}(t_0 + h)$.
Procedure

▷ Evaluate the right side of the differential equation six times,

$$\mathbf{f}_0 = \mathbf{f}(t_0, \mathbf{x}_0)$$

$$\mathbf{f}_\kappa = \mathbf{f}\left(t_0 + \alpha_\kappa h, \mathbf{x}_0 + h\sum_{\lambda=0}^{\kappa-1} \beta_{\kappa\lambda}\mathbf{f}_\lambda\right), \quad \kappa = 1, \ldots, 5.$$

▷ Compute the truncation error (*TE*) vector,

$$\Delta\mathbf{x} = h\sum_{\kappa=0}^{5}(c_\kappa - \hat{c}_\kappa)\mathbf{f}_\kappa.$$

▷ Find the maximum component of $\Delta\mathbf{x}$,

$$TEMAX = MAX\,|\Delta\mathbf{x}_\kappa|.$$

▷ Compute the fractional change in the step size,

$$\Delta = \left(\frac{TOL}{TEMAX + u} \right)^{\frac{1}{5}}.$$

▷ Test for a rejected or accepted step: if $TEMAX > TOL$, then the step has been rejected. Go back to the first step in this algorithm and repeat the computation with a smaller step size, which is the greater of $(\Delta)\, h$ or $(0.1)\, h$ (i.e., $h = h * MAX(\Delta, 0.1)$).

▷ If $TEMAX \leq TOL$, then the step is accepted. Compute the final solution step size $h_1 = h * MIN(\Delta, 4.0)$ for this step and the solution using either

$$\hat{\mathbf{x}}(t_0 + h_1) = \mathbf{x}_0 + h_1 \sum_{\kappa=0}^{5} \hat{c}_\kappa \mathbf{f}_\kappa \; \text{(higher} - \text{order solution)}$$

or

$$\mathbf{x}(t_0 + h_1) = \mathbf{x}_0 + h_1 \sum_{\kappa=0}^{4} c_\kappa \mathbf{f}_\kappa \; \text{(lower} - \text{order solution)}.$$

End.

Table (10.1) lists the $RKF4(5)$ coefficients taken from Fehlberg [32]. From this table, the TE formula is (using eq. 10.13)

$$TE = h \sum_{\kappa=0}^{5} (c_\kappa - \hat{c}_\kappa) f_\kappa$$

$$= h \left(-\frac{1}{360} f_0 + \frac{128}{4275} f_2 + \frac{2197}{75240} f_3 - \frac{1}{50} f_4 - \frac{2}{55} f_5 \right).$$

The Fehlberg reference [32] is to a paper that has an English version and a German version. Our colleague Robert Gottlieb discovered a discrepancy between the two versions. In the English paper the f_3 coefficient in the equation above for TE is wrong. In the German-language version the equation for TE is correct and is the same as the equation above.

Comments

- Note that in the $RKF4(5)$ algorithm the user may choose either the higher (fifth)-order or the lower (fourth)-order solution. A similar choice exists for any of the $RKFp(p+1)$ methods. Fehlberg designates the order of his methods by the lower number. For example, he refers to the $RKF4(5)$ method as a fourth-order method and chooses the fourth-order solution. In

Table 10.1 *RKF*4(5) Coefficients.

κ	α_κ	$\beta_{\kappa,0}$	$\beta_{\kappa,1}$	$\beta_{\kappa,2}$	$\beta_{\kappa,3}$	$\beta_{\kappa,4}$	c_κ	$\hat{c}_\kappa$
0	0	0					$\frac{25}{216}$	$\frac{16}{135}$
1	$\frac{1}{4}$	$\frac{1}{4}$					0	0
2	$\frac{3}{8}$	$\frac{3}{32}$	$\frac{9}{32}$				$\frac{1408}{2565}$	$\frac{6656}{12825}$
3	$\frac{12}{13}$	$\frac{1932}{2197}$	$-\frac{7200}{2197}$	$\frac{7296}{2197}$			$\frac{2197}{4104}$	$\frac{28561}{56430}$
4	1	$\frac{439}{216}$	-8	$\frac{3680}{513}$	$-\frac{845}{4104}$		$-\frac{1}{5}$	$-\frac{9}{50}$
5	$\frac{1}{2}$	$-\frac{8}{27}$	2	$-\frac{3544}{2565}$	$\frac{1859}{4104}$	$-\frac{11}{40}$		$\frac{2}{55}$

practice, most users would choose the fifth-order solution. Their justification is that since the higher-order solution is available with no additional function evaluations (that is, with no additional evaluations of the right side of the differential equations), it makes sense to use the higher-order solution. In the examples presented in chapter 9, the higher-order solutions were selected.

- Since several free parameters occur in the process of developing the equations of conditions for the *RKF*4(5), the table of coefficients given above are not unique. In fact, Fehlberg gives two tables of coefficients for the *RKF*4(5) in reference [32].

- At the point in the algorithm where the fractional change in the step size (Δ) is computed, we augment *TEMAX* by the addition of a small quantity (u) which is the *unit round-off error* of the computer [60], which is dependent on the computer being used. It is defined in computer arithmetic as the smallest number possible for which

$$1 + u > 1.$$

For many computers that use 64-bit IEEE floating-point arithmetic (often referred to as *double-precision*),

$$u \approx 2.2 \times 10^{-16}.$$

The addition of u to *TEMAX* prevents the denominator of Δ (*TEMAX* + u) from becoming zero. Note from the formula for *TE* that *TE* approaches zero as the functions f_0, f_2, f_3, f_4, and f_5 simultaneously approach the same value. That is,

$$\frac{-1}{360} + \frac{128}{4275} + \frac{2197}{75240} - \frac{1}{50} - \frac{2}{55} = 0.35 \times 10^{-17}.$$

11

TYPES OF PERTURBATIONS

Up to this point we have introduced certain perturbations in order to demonstrate the solution of narrowly defined examples. Several of these examples were given in chapter 8. We can group perturbations under two general classifications: those that arise from potential functions, and those that are not derivable from a potential function. In chapter 9, where we presented the Sperling-Burdet approach, each of these two classes was given its own place in the total perturbation function. That is,

$$\mathbf{F} = -\frac{\partial V}{\partial \mathbf{r}} + \mathbf{P},$$

where V is for a perturbing potential and $\mathbf{P}$ is for the other types of perturbations. Let us now discuss these classes separately. For brevity we follow our convention that m_1 is the more massive body and m_2 is a satellite either natural or artificial.

Perturbations arising from potentials are gravitational in nature. We list several of these as follows:

1. Third-body perturbations arise from other massive bodies such as the planets, Sun, and Moon. The potentials of third bodies are always explicitly time dependent since we can evaluate their effects only by knowing their positions as a function of time.

2. Planetary potential functions (the geopotential, for example) arise from the fact that m_1 is nonspherical, nonhomogeneous, and quite possibly rotating about some axis. In other words, m_1 is not a mass point or a finite homogeneous sphere (see §2.2.1), as tacitly assumed during the development of the two-body problem. Planetary potentials are explicitly, or implicitly, dependent on time (an explicit dependence arises due to rotation).

3. Tidal potential perturbations are a combination of third-body perturbations and planetary potential perturbations. This perturbation can best be

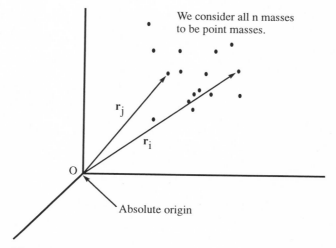

We consider all n masses
to be point masses.

$\mathbf{r}_j$

$\mathbf{r}_i$

O

Absolute origin

Figure 11.1   The n-body system.

understood by the example of the Moon (m_3) revolving about the Earth (m_1) and the effect on a geocentric satellite (m_2). As the Moon orbits the Earth it distorts the Earth's shape. This distortion in turn has a perturbing effect on the orbit of m_2.

Perturbations that do not arise from potentials are usually due to some "contact" accelerations acting on m_2. For example:

4. Molecules and other particles in the atmosphere impinging on satellite surfaces causing it to lose energy (drag acceleration).
5. The thrust of a rocket engine.

11.1 Third-Body Perturbations

Assume a system of n point masses (fig. 11.1). The equation of motion of the ith body with respect to an inertial system centered at an *absolute* origin is

$$m_i \ddot{\mathbf{r}}_i = -G m_i \sum_{\substack{j=1 \\ j \neq i}}^{n} \frac{m_j}{r_{ij}^3} (\mathbf{r}_i - \mathbf{r}_j),$$

where

$$i = 1, 2, \ldots, n$$

$$j = 1, 2, \ldots, n$$

$$r_{ij} = r_{ji} = \left\| \mathbf{r}_i - \mathbf{r}_j \right\|$$

$$\mathbf{r}_{ij} = -\mathbf{r}_{ji}.$$

Adding all of the equations of motion results in

$$\sum_{i}^{n} m_i \ddot{\mathbf{r}}_i = 0,$$

since for each $\mathbf{r}_{ij}$ term there is a $\mathbf{r}_{ji}$ term. This equation also states that the acceleration of the center of mass of an inertial system (a system in which Newton's laws apply) is zero.

We can integrate the above equation to get

$$\sum_{i}^{n} m_i \mathbf{r}_i = \mathbf{a}t + \mathbf{b},$$

where $\mathbf{a}$ and $\mathbf{b}$ are constants of integration. Since the center of mass of the n particle system is given by

$$\mathbf{r}_{cm} = \frac{\sum_{i=1}^{n} m_i \mathbf{r}_i}{\sum_{i=1}^{n} m_i},$$

we have

$$\mathbf{r}_{cm} = \frac{\mathbf{a}t + \mathbf{b}}{\sum_{i=1}^{n} m_i},$$

so the center of mass moves in a straight line with constant velocity (the sum of the accelerations is zero). The requirement for an absolute origin is that it *cannot be accelerated.* We can therefore place the center of mass of the system defined above at the absolute origin without changing the form of the equations of motion.

11.1.1 Development of the Perturbation

The differential equations of motion for the first three bodies of the n-body problem are

$$\ddot{\mathbf{r}}_1 = Gm_2 \frac{\mathbf{r}_2 - \mathbf{r}_1}{r_{21}^3} + Gm_3 \frac{\mathbf{r}_3 - \mathbf{r}_1}{r_{31}^3} + \mathbf{P}_1 \qquad (11.1)$$

$$\ddot{\mathbf{r}}_2 = Gm_1 \frac{\mathbf{r}_1 - \mathbf{r}_2}{r_{21}^3} + Gm_3 \frac{\mathbf{r}_3 - \mathbf{r}_2}{r_{23}^3} + \mathbf{P}_2 \qquad (11.2)$$

$$\ddot{\mathbf{r}}_3 = Gm_1 \frac{\mathbf{r}_1 - \mathbf{r}_3}{r_{31}^3} + Gm_2 \frac{\mathbf{r}_2 - \mathbf{r}_3}{r_{23}^3} + \mathbf{P}_3, \qquad (11.3)$$

where

$$\mathbf{P}_k = G \sum_{j=4}^{n} m_j \frac{\mathbf{r}_j - \mathbf{r}_k}{r_{jk}^3} \qquad (11.4)$$

and

$$k = 1, 2, 3$$

$$r_{lk} = r_{kl} = \|\mathbf{r}_l - \mathbf{r}_k\|$$

$$\mathbf{r}_{lk} = \mathbf{r}_l - \mathbf{r}_k.$$

Now identify the *first* body of the system as a planet, the *second* body as a spacecraft, and the *third* body as the Sun. Since all vectors are measured with respect to the absolute origin, $\mathbf{r}_1$, $\mathbf{r}_2$, and $\mathbf{r}_3$ are the positions of the planet, spacecraft, and Sun with respect to the *fixed* origin.

Now let us define the perturbed two-body problem of the motion of m_2 (the spacecraft) with respect to m_1 (the planet) by introducing the relative position vector

$$\mathbf{r} = \mathbf{r}_2 - \mathbf{r}_1. \qquad (11.5)$$

To find the differential equation for $\mathbf{r}$, subtract equation (11.1) from equation (11.2),

$$\ddot{\mathbf{r}}_2 - \ddot{\mathbf{r}}_1 = -Gm_1 \frac{\mathbf{r}_2 - \mathbf{r}_1}{r_{21}^3} + Gm_3 \frac{\mathbf{r}_3 - \mathbf{r}_2}{r_{23}^3} + \mathbf{P}_2$$

$$- Gm_2 \frac{\mathbf{r}_2 - \mathbf{r}_1}{r_{21}^3} - Gm_3 \frac{\mathbf{r}_3 - \mathbf{r}_1}{r_{31}^3} - \mathbf{P}_1,$$

which, after using equation (11.5), reduces to

$$\ddot{\mathbf{r}} = -G\frac{(m_1 + m_2)}{r^3}\mathbf{r} + Gm_3 \left(\frac{\mathbf{r}_3 - \mathbf{r}_2}{r_{23}^3} - \frac{\mathbf{r}_3 - \mathbf{r}_1}{r_{31}^3} \right) + \mathbf{P}_2 - \mathbf{P}_1. \qquad (11.6)$$

We remove the position of the spacecraft $\mathbf{r}_2$ from the right side of equation (11.6). Note that

$$\mathbf{r}_j - \mathbf{r}_2 = \mathbf{r}_j - \mathbf{r}_1 - \mathbf{r}_2 + \mathbf{r}_1 = (\mathbf{r}_j - \mathbf{r}_1) - \mathbf{r},$$

and so the magnitude is

$$\|\mathbf{r}_j - \mathbf{r}_2\| = \|(\mathbf{r}_j - \mathbf{r}_1) - \mathbf{r}\|.$$

Substitute these results in equation (11.6) to obtain

$$\ddot{\mathbf{r}} = -G\frac{(m_1 + m_2)}{r^3}\mathbf{r}$$
$$+ Gm_3 \left(\frac{(\mathbf{r}_3 - \mathbf{r}_1) - \mathbf{r}}{\|(\mathbf{r}_3 - \mathbf{r}_1) - \mathbf{r}\|^3} - \frac{\mathbf{r}_3 - \mathbf{r}_1}{r_{31}^3} \right) + \mathbf{P}_2 - \mathbf{P}_1, \qquad (11.7)$$

where

$$\mathbf{P}_2 = G \sum_{j=4}^{n} m_j \frac{(\mathbf{r}_j - \mathbf{r}_1) - \mathbf{r}}{\|(\mathbf{r}_j - \mathbf{r}_1) - \mathbf{r}\|^3} \qquad (11.8)$$

$$\mathbf{P}_1 = G \sum_{j=4}^{n} m_j \frac{\mathbf{r}_j - \mathbf{r}_1}{\|\mathbf{r}_j - \mathbf{r}_1\|^3}. \qquad (11.9)$$

Note the form of equations (11.7) along with the auxiliary equation (11.8) and (11.9). Since the perturbations are expressed as differences between positions, the location of the center of mass of the n bodies is not relevant to the solution. This is an *important* point since the positions of the planets are given with respect to the Sun (*not* the center of mass of the solar system).

Now we shift notation slightly. In equations (11.7), (11.8), and (11.9) substitute

$$\boldsymbol{\rho}_j = \mathbf{r}_j - \mathbf{r}_1$$

and

$$\mathbf{d}_j = \mathbf{r} - (\mathbf{r}_j - \mathbf{r}_1) = \mathbf{r} - \boldsymbol{\rho}_j,$$

(see fig. 11.2) which results in (11.7), (11.8), and (11.9) becoming

$$\ddot{\mathbf{r}} = -G\frac{(m_1 + m_2)}{r^3}\mathbf{r} - Gm_3 \left(\frac{\mathbf{d}_3}{d_3^3} + \frac{\boldsymbol{\rho}_3}{\rho^3} \right) + \mathbf{P}_2 - \mathbf{P}_1 \qquad (11.10)$$

$$\mathbf{P}_2 = -G \sum_{j=4}^{n} m_j \frac{\mathbf{d}_j}{d_j^3} \qquad (11.11)$$

$$\mathbf{P}_1 = G \sum_{j=4}^{n} m_j \frac{\boldsymbol{\rho}_j}{\rho_j^3}. \qquad (11.12)$$

Note that

$$\mathbf{P}_2 - \mathbf{P}_1 = -G \sum_{j=4}^{n} m_j \left(\frac{\mathbf{d}_j}{d_j^3} + \frac{\boldsymbol{\rho}_j}{\rho_j^3} \right).$$

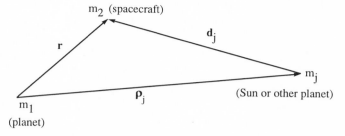

Figure 11.2 Position vectors in the *n*-body system.

Each term in the $\mathbf{P}_2 - \mathbf{P}_1$ equation is the *same in form* as the coefficient of m_3 in equation (11.10). In fact, we could include them all in the same summation.

11.1.2 Battin's Approach

Rewrite equation (11.10) as

$$\ddot{\mathbf{r}} + \frac{\mu}{r^3}\mathbf{r} = -G \sum_{j=3}^{n} m_j \left(\frac{1}{d_j^3}\mathbf{d}_j + \frac{1}{\rho_j^3}\boldsymbol{\rho}_j \right), \qquad (11.13)$$

where

$$\mu = G(m_1 + m_2).$$

Battin [5] introduces a reformulation of the right side of equation (11.13) which facilitates the solution when $r \ll \rho_j$ and thus $\mathbf{d}_j \approx -\boldsymbol{\rho}_j$. The term in equation (11.13),

$$\frac{1}{d_j^3}\mathbf{d}_j + \frac{1}{\rho_j^3}\boldsymbol{\rho}_j,$$

becomes inaccurate under this condition. Stiefel and Scheifele [64] call this the *interior problem*.

Battin [5] expresses equation (11.13) as

$$\ddot{\mathbf{r}} + \frac{\mu}{r^3}\mathbf{r} = -G \sum_{j=3}^{n} \frac{m_j}{d_j^3} \left(\mathbf{r} + f(q_j)\boldsymbol{\rho}_j \right), \qquad (11.14)$$

where

$$f(q_j) = q_j \left(\frac{3 + 3q_j + q_j^2}{1 + (1 + q_j)^{\frac{3}{2}}} \right)$$

$$q_j = \frac{1}{\rho_j^2} \mathbf{r} \cdot (\mathbf{r} - 2\rho_j).$$

This reformulation will be *proved* on the next few pages.

11.1.3 Derivation of f(q)

Consider the factor

$$\frac{\mathbf{d}_j}{d_j^3} + \frac{\rho_j}{\rho_j^3}.$$

The derivation applies to any term, so we drop the j subscript. Rewrite the above equation as

$$\frac{\mathbf{d}}{d^3} + \frac{\rho}{\rho^3} = \frac{1}{d^3}\left(\mathbf{d} + \frac{d^3}{\rho^3}\rho\right) = \frac{1}{d^3}\left(\underbrace{\mathbf{d} + \rho}_{\mathbf{r}} + \frac{d^3}{\rho^3}\rho - \rho\right),$$

which rearranges to

$$\frac{\mathbf{d}}{d^3} + \frac{\rho}{\rho^3} = \frac{1}{d^3}\left[\mathbf{r} + \left(\frac{d^3}{\rho^3} - 1\right)\rho\right]. \tag{11.15}$$

We want to find $(\frac{d^3}{\rho^3} - 1)$. Now,

$$\frac{d^2}{\rho^2} = \frac{(\mathbf{r} - \rho) \cdot (\mathbf{r} - \rho)}{\rho^2}$$

$$= \frac{1}{\rho^2}\mathbf{r} \cdot (\mathbf{r} - 2\rho) + 1.$$

Therefore, we can now introduce the defining relation for q,

$$q \equiv \frac{d^2}{\rho^2} - 1 = \frac{\mathbf{r} \cdot (\mathbf{r} - 2\rho)}{\rho \cdot \rho},$$

and we then have

$$\frac{d}{\rho} = \sqrt{q + 1}$$

and the defining equation for $f(q)$,

$$f(q) \equiv \frac{d^3}{\rho^3} - 1 = (q + 1)^{\frac{3}{2}} - 1. \tag{11.16}$$

Continuing with equation (11.15),

$$f(q) = \left((q+1)^{\frac{3}{2}} - 1\right) \left[\frac{(q+1)^{\frac{3}{2}} + 1}{(q+1)^{\frac{3}{2}} + 1} \right]$$

$$= \frac{(q+1)^3 - 1}{(q+1)^{3/2} + 1},$$

which becomes

$$f(q) = q \frac{q^2 + 3q + 3}{(q+1)^{\frac{3}{2}} + 1}. \tag{11.17}$$

For the *interior problem*,

$$q = \frac{d^2}{\rho^2} - 1 \approx 0,$$

since $d \approx \rho$. If in equation (11.17) we neglect q compared to 1, then

$$f(q) = q \frac{3 + \dots}{2 + \dots};$$

therefore,

$$f(q) \rightarrow \frac{3}{2}q,$$

as $q \rightarrow 0$ and the numerical problem in equation (11.13) is gone if we use the form (11.14).

11.1.4 Case When Position Is Large in Magnitude

The *exterior problem* [64] is also simplified by employing Battin's reformulation. If $r \gg \rho_3$, we see from figure 11.3 that $\mathbf{d}_3 \rightarrow \mathbf{r}$. Equation (11.13) then becomes

$$\ddot{\mathbf{r}} + \frac{\mu}{r^3}\mathbf{r} = -G\frac{m_3}{d_3^3}\left(\mathbf{r} + f(q_3)\rho_3\right) - G\sum_{j=4}^{n}\frac{m_j}{d_j^3}\left(\mathbf{r} + f(q_j)\rho_j\right), \tag{11.18}$$

where we have isolated the m_3 term from the summation.

We now develop an expression for $1/d_3^3$. From figure 11.3 we have

$$\mathbf{d}_3 = \mathbf{r} - \rho_3.$$

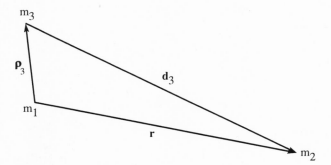

Figure 11.3   The exterior problem.

Calculate the magnitude of d_3 using the dot product,

$$\mathbf{d}_3 \cdot \mathbf{d}_3 = d_3^2 = r^2 + \rho_3^2 - 2\mathbf{r} \cdot \boldsymbol{\rho}_3,$$

which becomes

$$d_3 = r\sqrt{1 - 2\frac{\mathbf{r} \cdot \boldsymbol{\rho}_3}{r^2} + \frac{\rho_3^2}{r^2}},$$

and we therefore obtain

$$\frac{1}{d_3^3} = \frac{1}{r^3}\left(1 - 2\frac{\mathbf{r} \cdot \boldsymbol{\rho}_3}{r^2} + \frac{\rho_3^2}{r^2}\right)^{-\frac{3}{2}}.$$

We now define the quantities x and ν so that we can express the last equation in polynomials,

$$x = \frac{\rho_3}{r}$$

$$\nu = \hat{\mathbf{r}} \cdot \hat{\boldsymbol{\rho}}_3.$$

Substitute these definitions into the expression for $1/d_3^3$ to get

$$\frac{1}{d_3^3} = \frac{1}{r^3}\left(1 - 2\nu x + x^2\right)^{-\frac{3}{2}},$$

which can be expanded in an infinite series,

$$\frac{1}{d_3^3} = \frac{1}{r^3}\sum_{n=0}^{\infty} P_{n,3}(\nu)x^n,$$

where

$$P_{0,3} = 1$$

$$P_{1,3} = 3\nu$$

$$P_{2,3} = \frac{3}{2}\left(5\nu^2 - 1\right)$$

$$P_{3,3} = \frac{5}{2}\left(7\nu^3 - 3\nu\right)$$

$$\vdots$$

The expression for $1/d_3^3$ becomes

$$\frac{1}{d_3^3} = \frac{1}{r^3}\left(1 + \overbrace{\left[\sum_{n=1}^{\infty} P_{n,3}(\hat{\mathbf{r}}\cdot\hat{\boldsymbol{\rho}}_3)\left(\frac{\rho_3}{r}\right)^n\right]}^{\sigma}\right) = \frac{1}{r^3}(1+\sigma).$$

Now define σ to be the term in brackets and substitute $1/d_3^3$ into equation (11.18) to get

$$\ddot{\mathbf{r}} + \frac{G(m_1 + m_2)}{r^3}\mathbf{r} = -G\frac{m_3}{r^3}\left(\mathbf{r} + f(q_3)\boldsymbol{\rho}_3\right)(1+\sigma)$$

$$- G\sum_{j=4}^{n} \frac{m_j}{d_j^3}\left(\mathbf{r} + f(q_j)\boldsymbol{\rho}_j\right).$$

Collect terms in $G\frac{m_3}{r^3}\mathbf{r}$ on the left side and combine terms to obtain

$$\ddot{\mathbf{r}} + \frac{G(m_1 + m_2 + m_3)}{r^3}\mathbf{r} = -G\frac{m_3}{r^3}\left(\sigma\mathbf{r} + (1+\sigma)f(q_3)\boldsymbol{\rho}_3\right)$$

$$- G\sum_{j=4}^{n} \frac{m_j}{d_j^3}\left(\mathbf{r} + f(q_j)\boldsymbol{\rho}_j\right). \qquad (11.19)$$

For $r \gg \rho$, the problem approaches that of a perturbed two-body problem with m_1 and m_3 taken as one body. Note that part of the effect of m_3 also appears as a perturbation.

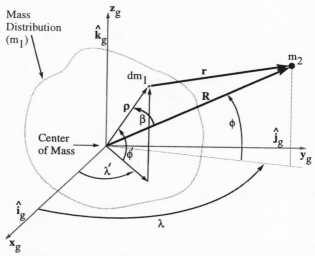

Figure 11.4 Generalized (nonspherical, nonhomogeneous) planetary mass distribution.

11.2 Potential Function for a Planet

Consider the potential of a mass (m_2) due to an arbitrary, nonhomogeneous and nonspherical mass (m_1) [53]. The mass m_2 is a point mass with coordinates (R, ϕ, λ) (refer to fig. 11.4). The mass element dm_1 is an increment of mass m_1 with coordinates (ρ, ϕ', λ'). The coordinate system x_g, y_g, z_g is fixed to the planet (*body-fixed*) and rotates at constant angular velocity. The gravitational potential energy is

$$U = -G\, m_2 \int_{m_1} \frac{dm_1}{r}. \tag{11.20}$$

Since $\mathbf{r} = \mathbf{R} - \boldsymbol{\rho}$ and $\mathbf{R} \cdot \boldsymbol{\rho} = R\, \rho \cos \beta$ we take the dot product of $\mathbf{r}$ with itself,

$$r^2 = \mathbf{r} \cdot \mathbf{r} = r^2 + \rho^2 - 2R\rho \cos \beta,$$

which becomes

$$r = R \left(1 - \frac{2\mathbf{R} \cdot \boldsymbol{\rho}}{R^2} + \frac{\rho^2}{R^2} \right)^{\frac{1}{2}}$$

$$\frac{1}{r} = \frac{1}{R}\left(1 - \frac{2\mathbf{R}\cdot\rho}{R^2} + \frac{\rho^2}{R^2}\right)^{-\frac{1}{2}}.$$

Following Arfken [1] we expand the right side of the last equation in Legendre polynomials ($P_n\,(\cos\beta)$) to get

$$\frac{1}{r} = \frac{1}{R}\sum_{n=0}^{\infty}\left(\frac{\rho}{R}\right)^n P_n\,(\cos\beta).$$

Substitute into equation (11.20) for the potential energy to obtain

$$U = \frac{-G\,m_2}{R}\sum_{n=0}^{\infty}\int_{m_1}\left(\frac{\rho}{R}\right)^n P_n\,(\cos\beta)\,dm_1. \qquad (11.21)$$

The Legendre polynomials are given by (letting $v = \cos\beta$)

$$P_0(v) = 1$$

$$P_1(v) = v$$

$$P_2(v) = \frac{1}{2}\left(3v^2 - 1\right)$$

$$\vdots$$

$$nP_n = (2n - 1)vP_{n-1}(v) - (n - 1)P_{n-2}(v).$$

Now evaluate $\cos\beta$ in terms of $\lambda, \lambda', \phi, \phi'$ directly from $R\rho\cos\beta = \mathbf{R}\cdot\rho$,

$$\cos\beta = \sin\phi\sin\phi' + \cos\phi\cos\phi'\cos(\lambda - \lambda'),$$

and substitute into equation (11.21). After considerable algebra,

$$U = -\frac{G\,m_1 m_2}{R}\left(1 + \sum_{n=1}^{\infty}\left[\left(\frac{a_e}{R}\right)^n C_{n,0}P_n\,(\sin\phi)\right.\right.$$

$$\left.\left. + \sum_{m=1}^{n}\left(\frac{a_e}{R}\right)^n P_{n,m}(\sin\phi)(C_{n,m}\cos m\lambda + S_{n,m}\sin m\lambda)\right]\right), \qquad (11.22)$$

where

$$a_e = \text{equatorial radius of the body } m_1.$$

Also,

$$C_{n,0} = \frac{1}{a_e^n\,m_1}\int_{m_1}\rho^n P_{n,m}(\sin\lambda')\,dm_1$$

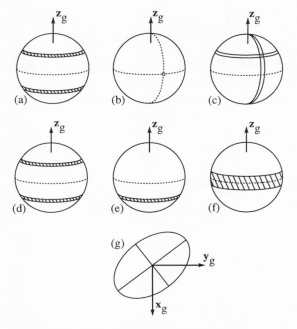

Figure 11.5 Mass distribution of planet relative to zonal, sectorial, and tesseral terms.

$$C_{n,m} = \frac{2}{a_e^n \, m_1} \frac{(n-m)!}{(n+m)!} \int_{m_1} \rho^n P_{n,m}(\sin \lambda') \cos \lambda' \, dm_1$$

$$S_{n,m} = \frac{2}{a_e^n \, m_1} \frac{(n-m)!}{(n+m)!} \int_{m_1} \rho^n P_{n,m}(\sin \lambda') \sin \lambda' \, dm_1.$$

The $C_{n,0}$, $C_{n,m}$, $S_{n,m}$ are determined by observation (tracking) of satellites. The $P_{n,m}(\nu)$ are called *associated Legendre functions* given by

$$P_{n,m}(\nu) = (1 - \nu^2)^{\frac{m}{2}} \frac{\partial^m P_n(\nu)}{\partial \nu^m},$$

where $P_n(\nu)$ are the Legendre polynomials.

Since the x_g, y_g, z_g axes have their origin at the center of mass, we can start the summation at $n = 2$. The planet rotates about the z_g-axis. The subscript n is called the *degree* of the model and the subscript m is called the *order* of the model. Note also that $m \leq n$. Figure 11.5 shows the mass distribution of the planet for some zonal, sectorial, and tesseral terms.

Terms having $m = 0$ ($C_n, 0$) are called *zonals* (a); these depend only on latitude. Terms having $n = m$ ($C_{n,n}$, $S_{n,n}$) are called *sectorials* (b); these depend only on longitude. *Tesseral* terms (c) depend on latitude and longitude ($n > m \neq 0$). Even zonals (d) (n even with $m = 0$) are symmetric with respect to the equator. Odd zonals (e) (n odd with $m = 0$) are not symmetric with respect to the equator. An example is the $C_{3,0}$ odd zonal, which indicates a "pear-shaped" Earth. The second even zonal (f) ($C_{2,0}$) is the oblateness term (symmetric excess mass distribution about the equator). This is the largest effect except for the $1/R$ term, which is the two-body term. The second even zonal ($C_{2,0}$) and the second sectorials ($C_{2,2}$ and $S_{2,2}$) (g) define the triaxial ellipsoid (equatorial ellipticity). Note that in figure 11.5g the z_g axis is out of the page.

11.2.1 Case of a Satellite about the Earth

The "geopotential" function is

$$V_T = \frac{U}{m_2},$$

where m_2 is the mass of the satellite, m_1 is the mass of the Earth, so $Gm_1 = GE$; $r = R$ (note the change in notation) is the distance of the satellite from the center of mass.

The geopotential function (from eq. 11.22) becomes

$$V_T = -\frac{GE}{r} - \frac{GE}{r} \sum_{n=2}^{\infty} \left[\left(\frac{a_e}{r}\right)^n \sum_{m=0}^{n} (S_{n,m} \sin m\lambda \right.$$

$$\left. + C_{n,m} \cos m\lambda) P_{n,m}(\sin \phi) \right], \tag{11.23}$$

where the first term is the two-body potential and the remainder is the perturbing potential V. The equation of motion of the satellite is

$$\ddot{\mathbf{r}} = -\frac{\partial V_T}{\partial \mathbf{r}} = GE \frac{\partial}{\partial \mathbf{r}} \left(\frac{1}{r}\right) - \frac{\partial V}{\partial \mathbf{r}}.$$

Consider the gradient of the first term of equation (11.23) (which is the two-body potential),

$$\frac{\partial}{\partial \mathbf{r}} \left(\frac{1}{r}\right) = \frac{\partial}{\partial \mathbf{r}} (\mathbf{r} \cdot \mathbf{r})^{-\frac{1}{2}} = -\frac{1}{2} (\mathbf{r} \cdot \mathbf{r})^{-\frac{3}{2}} (2\mathbf{r}^T) \frac{\partial \mathbf{r}}{\partial \mathbf{r}}$$

$$= -\frac{1}{r^3} \mathbf{r}.$$

So equation for $\ddot{\mathbf{r}}$ now becomes

$$\ddot{\mathbf{r}} + \frac{G\,E}{r^3}\mathbf{r} = -\frac{\partial V}{\partial \mathbf{r}},$$

which is an equation of perturbed two-body motion. Since m_2 (mass of the satellite) is negligible, the conventional μ becomes

$$\mu = G\,(m_1 + m_2) = G\,m_1 = G\,E.$$

The part of equation (11.23) which is the perturbing potential is

$$V = -\frac{G\,E}{r} \sum_{n=2}^{\infty} \left[\left(\frac{a_e}{r}\right)^n \sum_{m=0}^{n} (S_{n,m} \sin m\lambda \right.$$

$$\left. + C_{n,m} \cos m\lambda) P_{n,m}(\sin\phi) \right]. \qquad (11.24)$$

The development of the gradient of this function is straightforward, though laborious. The major problem, however, is a computational one. The number of trigonometric functions that must be evaluated is very large. This problem was solved by Mueller [55] in 1975 and Gottlieb [36] in 1993 by the development of a recursive, nonsingular (except for $r = 0$) algorithm for V and $\partial V / \partial \mathbf{r}$. We present the development without proof. The potential is

$$V = -\sum_{n=2}^{\infty} \sum_{m=0}^{n} V_{n,m},$$

and its gradient in the Earth-fixed system is

$$\frac{\partial V}{\partial \mathbf{r}_g} = -\sum_{n=2}^{\infty} \sum_{m=0}^{n} \frac{\partial V_{n,m}}{\partial \mathbf{r}_g},$$

where the individual terms in the summations are

$$V_{n,m} = \frac{G\,E}{r} \left(\frac{a_e}{r}\right)^n \frac{B_{n,m}}{n+m+1}$$

and

$$\frac{\partial V_{n,m}}{\partial \mathbf{r}_g} = G\,E \left(\frac{a_e}{r}\right)^n \frac{1}{r^2} \left[-\hat{\mathbf{r}}_g (\nu H_{n,m} + B_{n,m}) \right.$$

$$\left. + \hat{\mathbf{i}}_g D_{n,m} - \hat{\mathbf{j}}_g E_{n,m} + \hat{\mathbf{k}}_g H_{n,m} \right].$$

The functions $B_{n,m}$, $E_{n,m}$, $D_{n,m}$ and $H_{n,m}$ are

$$B_{n,m} = (C_{n,m}\tilde{C}_m + S_{n,m}\tilde{S}_m)(n+m+1)P_n^m$$

$$E_{n,m} = -m(C_{n,m}\tilde{S}_{m-1} - S_{n,m}\tilde{C}_{m-1})P_n^m$$

$$D_{n,m} = m(C_{n,m}\tilde{C}_{m-1} + S_{n,m}\tilde{S}_{m-1})P_n^m$$

$$H_{n,m} = (C_{n,m}\tilde{C}_m + S_{n,m}\tilde{S}_m)P_n^{m+1}.$$

The functions labeled with a $\sim$ are

$$\tilde{C}_m = \tilde{C}_1\tilde{C}_{m-1} - \tilde{S}_1\tilde{S}_{m-1}$$

$$\tilde{S}_m = \tilde{S}_1\tilde{C}_{m-1} + \tilde{C}_1\tilde{S}_{m-1},$$

where the starting values are

$$\tilde{C}_0 = 1 \qquad \tilde{S}_0 = 0$$

$$\tilde{C}_1 = x_g/r \quad \tilde{S}_1 = y_g/r.$$

The functions P_n^m are the derivatives of the Legendre polynomials and are given by

$$P_n^0 = \frac{1}{n}\left[(2n-1)vP_{n-1}^0 - (n-1)P_{n-2}^0\right]; \quad m = 0$$

and

$$P_n^m = P_{n-2}^m + (2n-1)P_{n-1}^{m-1}; \quad n \geq 2 \text{ and } m \leq n,$$

where the starting values are given by

$$P_0^0 = 1 \quad P_1^0 = v = z_g/r$$

$$P_0^1 = 0 \quad P_1^1 = 1.$$

In this algorithm the gradient $\partial V/\partial \mathbf{r}_g$ is found in an Earth-fixed coordinate system. Since the gradient of the potential normally must be given in an inertial system, we must define the transformation between the inertial and Earth-fixed systems. An inertial coordinate system can be conveniently defined placing the origin at the center of the Earth, specifying the fundamental plane to be the mean Earth equator, and specifying the principal direction to be the intersection of the mean equator and the Greenwich Meridian (the Earth's prime meridian) at midnight (UT1 = 0; see definition of universal time in §1.3). This inertial coordinate system (x, y, z) and the Earth-fixed system (x_g, y_g, z_g) are shown in figure 8.4. The transformation between these two coordinate systems is given by

$$
\begin{pmatrix} x_g \\ y_g \\ z_g \end{pmatrix} = \begin{bmatrix} \cos \omega t & \sin \omega t & 0 \\ -\sin \omega t & \cos \omega t & 0 \\ 0 & 0 & 1 \end{bmatrix} \begin{pmatrix} x \\ y \\ z \end{pmatrix}
$$

from Appendix A. See figure A.3, which illustrates a rotation about the x_3 ($= z_g$) axis. Refer to §1.3 for the angular velocity (ω) of the Earth.

APPENDIXES

A

COORDINATE TRANSFORMATIONS

Here we consider the transformation of a vector from one cartesian coordinate system to another by the use of orthogonal matrices. An *orthogonal matrix* is a matrix with an inverse equal to its transpose, that is, if A is an orthogonal matrix,

$$A^{-1} = A^T$$

$$AA^T = A^T A = I.$$

Clearly implicit here is that an orthogonal matrix is square (equal number of rows and columns) and is invertible.

An orthogonal matrix is unitary—the magnitude (norm) of a vector is invariant under transformation by an orthogonal matrix. This is easy to prove:[1]

$$\mathbf{x} = A\mathbf{y}$$

$$\langle \mathbf{x}, \mathbf{x} \rangle = \langle A\mathbf{y}, A\mathbf{y} \rangle$$

$$\|\mathbf{x}\|^2 = \langle A^T A\mathbf{y}, \mathbf{y} \rangle$$

$$= \langle I\mathbf{y}, \mathbf{y} \rangle$$

$$\|\mathbf{x}\|^2 = \|\mathbf{y}\|^2$$

A.1 Rotation of Coordinate Systems

In this section we develop the equations for rotating a vector about each of the three cartesian axes. Refer to figures A.1–A.3.

---

[1] The $\langle \mathbf{x}, \mathbf{y} \rangle$ notation is a standard notation for the *inner* (or scalar) product. See reference [62], page 208.

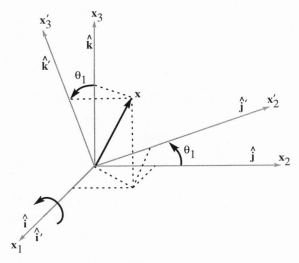

Figure A.1 Rotation about the x_1-axis.

A positive (counterclockwise) rotation about the x_1-axis through an angle θ_1 produces the following relation among the components of a vector $(\mathbf{x})$ in the *old* system (x_1, x_2, x_3) and the *new* system (x_1', x_2', x_3'):

$$\begin{pmatrix} x_1' \\ x_2' \\ x_3' \end{pmatrix} = \begin{bmatrix} 1 & 0 & 0 \\ 0 & \cos\theta_1 & \sin\theta_1 \\ 0 & -\sin\theta_1 & \cos\theta_1 \end{bmatrix} \begin{pmatrix} x_1 \\ x_2 \\ x_3 \end{pmatrix}.$$

This equation can be written concisely as

$$\mathbf{x}' = A(\theta_1)\mathbf{x}.$$

The form of the matrix $A(\theta_1)$ is derived by writing the vector equation for $\mathbf{x}$ in the x_1', x_2', x_3' coordinate system in terms of the components in the x_1, x_2, x_3 system. We start with

$$\mathbf{x} = \hat{\mathbf{i}}'x_1' + \hat{\mathbf{j}}'x_2' + \hat{\mathbf{k}}'x_3' = \hat{\mathbf{i}}x_1 + \hat{\mathbf{j}}x_2 + \hat{\mathbf{k}}x_3.$$

We take the dot product of $\mathbf{x}$ in the above equation with the unit vectors $\hat{\mathbf{i}}'$, $\hat{\mathbf{j}}'$, $\hat{\mathbf{k}}'$ to obtain the desired relationships,

$$\hat{\mathbf{i}}' \cdot \mathbf{x} = x_1' = (\hat{\mathbf{i}}' \cdot \hat{\mathbf{i}})x_1 + (\hat{\mathbf{i}}' \cdot \hat{\mathbf{j}})x_2 + (\hat{\mathbf{i}}' \cdot \hat{\mathbf{k}})x_3$$

$$\hat{\mathbf{j}}' \cdot \mathbf{x} = x_2' = (\hat{\mathbf{j}}' \cdot \hat{\mathbf{i}})x_1 + (\hat{\mathbf{j}}' \cdot \hat{\mathbf{j}})x_2 + (\hat{\mathbf{j}}' \cdot \hat{\mathbf{k}})x_3$$

$$\hat{\mathbf{k}}' \cdot \mathbf{x} = x_3' = (\hat{\mathbf{k}}' \cdot \hat{\mathbf{i}})x_1 + (\hat{\mathbf{k}}' \cdot \hat{\mathbf{j}})x_2 + (\hat{\mathbf{k}}' \cdot \hat{\mathbf{k}})x_3,$$

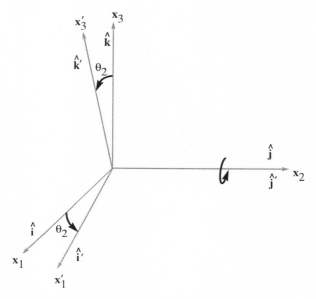

Figure A.2   Rotation about the x_2-axis.

or in matrix form,

$$\begin{pmatrix} x_1' \\ x_2' \\ x_3' \end{pmatrix} = \begin{bmatrix} \hat{\mathbf{i}}' \cdot \hat{\mathbf{i}} & \hat{\mathbf{i}}' \cdot \hat{\mathbf{j}} & \hat{\mathbf{i}}' \cdot \hat{\mathbf{k}} \\ \hat{\mathbf{j}}' \cdot \hat{\mathbf{i}} & \hat{\mathbf{j}}' \cdot \hat{\mathbf{j}} & \hat{\mathbf{j}}' \cdot \hat{\mathbf{k}} \\ \hat{\mathbf{k}}' \cdot \hat{\mathbf{i}} & \hat{\mathbf{k}}' \cdot \hat{\mathbf{j}} & \hat{\mathbf{k}}' \cdot \hat{\mathbf{k}} \end{bmatrix} \begin{pmatrix} x_1 \\ x_2 \\ x_3 \end{pmatrix}.$$

The nine dot products in the above matrix are called *direction cosines*, and can be worked out by referring to figure A.1. For example, the $\hat{\mathbf{k}}' \cdot \hat{\mathbf{j}}$ term is

$$\hat{\mathbf{k}}' \cdot \hat{\mathbf{j}} = \cos \left(\frac{\pi}{2} + \theta_1 \right) = -\sin \theta_1.$$

Substitution of the various direction cosines recovers the matrix $A(\theta_1)$.

We next consider a positive rotation about the x_2-axis, diagrammed in figure A.2. Using this figure we work out the direction cosines to obtain

$$\begin{pmatrix} x_1' \\ x_2' \\ x_3' \end{pmatrix} = \begin{bmatrix} \cos \theta_2 & 0 & -\sin \theta_2 \\ 0 & 1 & 0 \\ \sin \theta_2 & 0 & \cos \theta_2 \end{bmatrix} \begin{pmatrix} x_1 \\ x_2 \\ x_3 \end{pmatrix},$$

which we write in concise form as

$$\mathbf{x}' = B(\theta_2)\mathbf{x}.$$

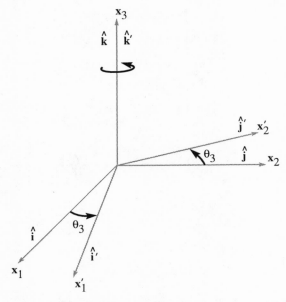

Figure A.3   Rotation about the x_3-axis.

Finally we consider a positive rotation about the x_3-axis, diagrammed in figure A.3. Using the same procedure as above results in

$$\begin{pmatrix} x'_1 \\ x'_2 \\ x'_3 \end{pmatrix} = \begin{bmatrix} \cos\theta_3 & \sin\theta_3 & 0 \\ -\sin\theta_3 & \cos\theta_3 & 0 \\ 0 & 0 & 1 \end{bmatrix} \begin{pmatrix} x_1 \\ x_2 \\ x_3 \end{pmatrix},$$

which we write in concise form as

$$\mathbf{x}' = C(\theta_3)\mathbf{x}.$$

A.2 Transformation to the Two-Body System

An orthogonal system is defined by an origin and three mutually perpendicular axes. An inertial cartesian coordinate system is orthogonal by definition. Inertial cartesian systems are defined by specifying an origin, a fundamental plane, and a principal direction contained in the fundamental plane. The three mutually perpendicular axes are the principal direction, an axis perpendicular to the fundamental plane, and a third axis defined as the cross product of the

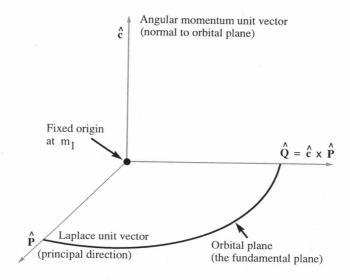

Figure A.4   An orthogonal inertial (two-body $\hat{\mathbf{P}}$, $\hat{\mathbf{Q}}$, $\hat{\mathbf{c}}$) system.

other two axes following the right-hand rule. For the two-body problem we define the origin at m_1, the fundamental plane to be the plane of the orbit, and the principal direction to be the Laplace vector. This is the $\hat{\mathbf{P}}$, $\hat{\mathbf{Q}}$, $\hat{\mathbf{c}}$ coordinate system (see fig. A.4).

The $\hat{\mathbf{P}}$, $\hat{\mathbf{Q}}$, $\hat{\mathbf{c}}$ system is a convenient system but in a practical sense is not very interesting, since there is a different $\hat{\mathbf{P}}$, $\hat{\mathbf{Q}}$, $\hat{\mathbf{c}}$ system for every two-body orbit, even if we keep the origin fixed at m_1. This system also suffers from the fact that it is not defined for circular orbits (which have no pericenter), nor is it defined for rectilinear orbits ($\mathbf{r}$ and $\dot{\mathbf{r}}$ are parallel, so $\mathbf{c}$ is undefined).

The $\hat{\mathbf{P}}$, $\hat{\mathbf{Q}}$, $\hat{\mathbf{c}}$ system, or any other system defined by two-body motion, can be related to an inertial cartesian system by (at most) three rotational transformations (we assume that the systems have a common origin with no loss of generality). The inertial cartesian system will be defined with respect to an origin, a fundamental plane, and a principal direction that is *independent* of any two-body motion.

We shall now assume that we have an arbitrary inertial system centered at m_1 with arbitrary fundamental plane and principal direction. The system is diagrammed in figure A.5.

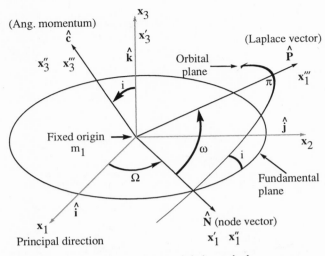

Figure A.5   Relationship between $\hat{\mathbf{i}}, \hat{\mathbf{j}}, \hat{\mathbf{k}}$ and $\hat{\mathbf{P}}, \hat{\mathbf{Q}}, \hat{\mathbf{c}}$ systems.

The transformation from the $\hat{\mathbf{i}}, \hat{\mathbf{j}}, \hat{\mathbf{k}}$ system to the $\hat{\mathbf{P}}, \hat{\mathbf{Q}}, \hat{\mathbf{c}}$ system is found from the three axis rotations we derived in the previous section. The first rotation is a positive (counterclockwise) rotation about the x_3-axis through the angle of the ascending node $\Omega \ (= \theta_3)$, resulting in

$$\mathbf{x}' = C(\Omega)\mathbf{x}. \tag{A.1}$$

Note that x_1', x_2', x_3' are along the direction $\hat{\mathbf{N}}, \hat{\mathbf{k}} \times \hat{\mathbf{N}}, \hat{\mathbf{k}}$, respectively. In fact, the unit vectors of the new system are

$$\hat{\mathbf{i}}' = \hat{\mathbf{N}}$$

$$\hat{\mathbf{j}}' = \hat{\mathbf{k}} \times \hat{\mathbf{N}}$$

$$\hat{\mathbf{k}}' = \hat{\mathbf{k}}.$$

The next rotation is a positive rotation about the x_1'-axis through the inclination angle $i \ (= \theta_1)$, resulting in

$$\mathbf{x}'' = A(i)\mathbf{x}'. \tag{A.2}$$

The x_1'', x_2'', x_3'' axes are along $\hat{\mathbf{N}}, \hat{\mathbf{c}} \times \hat{\mathbf{N}}, \hat{\mathbf{c}}$.

The final rotation is a positive rotation about the x_3''-axis through the angle ω

$(= \theta_3)$, which is called the *argument of pericenter*, resulting in

$$\mathbf{x}''' = C(\omega)\mathbf{x}'', \tag{A.3}$$

with the x_1''', x_2''', x_3''' axes along $\hat{\mathbf{P}}$, $\hat{\mathbf{Q}} = \hat{\mathbf{c}} \times \hat{\mathbf{P}}$, $\hat{\mathbf{c}}$.

We now consider each transformation in more detail. The *primed* system is found from (A.1):

$$\begin{pmatrix} x_1' \\ x_2' \\ x_3' \end{pmatrix} = \begin{bmatrix} \cos\Omega & \sin\Omega & 0 \\ -\sin\Omega & \cos\Omega & 0 \\ 0 & 0 & 1 \end{bmatrix} \begin{pmatrix} x_1 \\ x_2 \\ x_3 \end{pmatrix}.$$

The top row of this matrix gives us the components of the nodal vector $\hat{\mathbf{N}}$ (or the direction of x_1'),

$$\hat{\mathbf{N}} = \hat{\mathbf{i}}\cos\Omega + \hat{\mathbf{j}}\sin\Omega, \tag{A.4}$$

as can be easily seen from figure A.5.

The *double-primed* system is found from equations (A.1) and (A.2):

$$\mathbf{x}'' = A(i)C(\Omega)\mathbf{x},$$

or

$$\begin{pmatrix} x_1'' \\ x_2'' \\ x_3'' \end{pmatrix} = \begin{bmatrix} \cos\Omega & \sin\Omega & 0 \\ -\cos i \sin\Omega & \cos i \cos\Omega & \sin i \\ \sin i \sin\Omega & -\sin i \cos\Omega & \cos i \end{bmatrix} \begin{pmatrix} x_1 \\ x_2 \\ x_3 \end{pmatrix}.$$

The middle row of this matrix gives us

$$\hat{\mathbf{c}} \times \hat{\mathbf{N}} = -\hat{\mathbf{i}}\cos i \sin\Omega + \hat{\mathbf{j}}\cos i \cos\Omega + \hat{\mathbf{k}}\sin i, \tag{A.5}$$

and the bottom row gives us the angular momentum unit vector,

$$\hat{\mathbf{c}} = \hat{\mathbf{i}}\sin i \sin\Omega - \hat{\mathbf{j}}\sin i \cos\Omega + \hat{\mathbf{k}}\cos i. \tag{A.6}$$

The *triple-primed* system is found from equations (A.1), (A.2), and (A.3):

$$\mathbf{x}''' = C(\omega)A(i)C(\Omega)\mathbf{x},$$

or

$$\begin{pmatrix} x_1''' \\ x_2''' \\ x_3''' \end{pmatrix} = \begin{bmatrix} \cos\omega & \sin\omega & 0 \\ -\sin\omega & \cos\omega & 0 \\ 0 & 0 & 1 \end{bmatrix}$$

$$\times \begin{bmatrix} \cos\Omega & \sin\Omega & 0 \\ -\cos i \sin\Omega & \cos i \cos\Omega & \sin i \\ \sin i \sin\Omega & -\sin i \cos\Omega & \cos i \end{bmatrix} \begin{pmatrix} x_1 \\ x_2 \\ x_3 \end{pmatrix}. \quad (A.7)$$

The rows of the product of the two matrices in equation (A.7) are the unit vectors of the $\hat{\mathbf{P}}, \hat{\mathbf{Q}}, \hat{\mathbf{c}}$ system expressed in the $\hat{\mathbf{i}}, \hat{\mathbf{j}}, \hat{\mathbf{k}}$ system. These vectors are

$$\hat{\mathbf{P}} = \hat{\mathbf{i}}(\cos\omega\cos\Omega - \sin\omega\sin\Omega\cos i)$$

$$+ \hat{\mathbf{j}}(\cos\omega\sin\Omega + \sin\omega\cos\Omega\cos i)$$

$$+ \hat{\mathbf{k}}\sin\omega\sin i \quad (A.8)$$

$$\hat{\mathbf{Q}} = -\hat{\mathbf{i}}(\sin\omega\cos\Omega + \cos\omega\sin\Omega\cos i)$$

$$- \hat{\mathbf{j}}(\sin\omega\sin\Omega - \cos\omega\cos\Omega\cos i)$$

$$+ \hat{\mathbf{k}}\cos\omega\sin i \quad (A.9)$$

$$\hat{\mathbf{c}} = \hat{\mathbf{i}}\sin i\sin\Omega - \hat{\mathbf{j}}\sin i\cos\Omega + \hat{\mathbf{k}}\cos i. \quad (A.10)$$

The transformation matrix rotating the arbitrary inertial system $\hat{\mathbf{i}}, \hat{\mathbf{j}}, \hat{\mathbf{k}}$ to the $\hat{\mathbf{P}}, \hat{\mathbf{Q}}, \hat{\mathbf{c}}$ system can be written concisely as

$$M = C(\omega)A(i)C(\Omega) = \begin{bmatrix} \hat{\mathbf{P}}^T \\ \hat{\mathbf{Q}}^T \\ \hat{\mathbf{c}}^T \end{bmatrix}, \quad (A.11)$$

that is, equation (A.7) can be written

$$\mathbf{x}''' = M\mathbf{x}. \quad (A.12)$$

B

HYPERBOLIC MOTION

Recall from §2.4.7, when we were discussing the development of Kepler's equation (the final integral) from the angular momentum integral,

$$\mathbf{c} = \mathbf{r} \times \dot{\mathbf{r}} = constant,$$

we developed the equation

$$c^2 = r^2 \left(2h + \frac{2\mu}{r} \right) - r^2 \dot{r}^2. \tag{B.1}$$

We now develop Kepler's equation for the case where the energy is *positive*, $h > 0$. We start by solving equation (B.1) for $r\,\dot{r}$:

$$r^2 \dot{r}^2 = 2hr^2 + 2\mu r - c^2.$$

Factor $(2h)$ and take the square root of both sides:

$$r\,\dot{r} = \sqrt{2h} \sqrt{r^2 + \frac{\mu}{h} r - \frac{c^2}{2h}}. \tag{B.2}$$

Recall that previously we had factored $\sqrt{-2h}$, which is suitable for $h < 0$.

Now we continue the development by reducing equation (B.2) to the form

$$dt = f(r)\,dr,$$

which we will integrate. From equation (B.2),

$$r\,\dot{r} = \sqrt{2h} \sqrt{\left(r + \frac{\mu}{2h} \right)^2 - \left(\frac{\mu}{2h} \right)^2 - \frac{c^2}{2h}}. \tag{B.3}$$

But recall from equation (2.23) that

$$\frac{c^2}{\mu} = -\frac{\mu}{2h}\left(1 - \left(\frac{P}{\mu}\right)^2\right),$$

which we multiply by $\mu/2h$ to get

$$\frac{c^2}{2h} = -\left(\frac{\mu}{2h}\right)^2\left(1 - \left(\frac{P}{\mu}\right)^2\right)$$

or

$$\frac{c^2}{2h} + \left(\frac{\mu}{2h}\right)^2 = \left(\frac{P}{2h}\right)^2,$$

which we substitute on the right side of equation (B.3) to obtain

$$r\frac{dr}{dt} = \sqrt{2h}\sqrt{\left(r + \frac{\mu}{2h}\right)^2 - \left(\frac{P}{2h}\right)^2},$$

which becomes

$$\sqrt{2h}\,dt = \frac{r\,dr}{\sqrt{\left(r + \frac{\mu}{2h}\right)^2 - \left(\frac{P}{2h}\right)^2}}. \qquad (B.4)$$

Compare equation (B.4) with equation (2.31).

We now change variables to simplify equation (B.4) to a form that appears in a table of integrals. Define

$$z = r + \frac{\mu}{2h}, \quad dz = dr. \qquad (B.5)$$

Substituting into equation (B.4), we get

$$\sqrt{2h}\int dt = \int \frac{z\,dz}{\sqrt{\left[z^2 - \left(\frac{P}{2h}\right)^2\right]}} - \frac{\mu}{2h}\int \frac{dz}{\sqrt{\left[z^2 - \left(\frac{P}{2h}\right)^2\right]}}. \qquad (B.6)$$

Now integrate to obtain

$$\sqrt{2h}\,t + constant = \sqrt{z^2 - \left(\frac{P}{2h}\right)^2}$$

$$-\frac{\mu}{2h}\sinh^{-1}\left(\frac{\sqrt{z^2 - \left(\frac{P}{2h}\right)^2}}{P/2h}\right). \qquad (B.7)$$

Define the quantity

$$\sinh H = \frac{2h}{P}\sqrt{z^2 - \left(\frac{P}{2h}\right)^2}.$$
(B.8)

Using this definition in equation (B.7) results in

$$\sqrt{2h}\, t + \text{constant} = \frac{P}{2h}\sinh H - \frac{\mu}{2h}H.$$

Multiply through by $2h/\mu$ to obtain

$$\frac{2h}{\mu}\sqrt{2h}\,(t+K) = \frac{P}{\mu}\sinh H - H,$$
(B.9)

where K is the constant of integration. Equation (B.9) is the hyperbolic equivalent of *Kepler's equation*.

We can also get an equation for the distance r. From equation (B.8) and the relation $\cosh^2 H - \sinh^2 H = 1$,

$$\cosh^2 H = 1 + \sinh^2 H = 1 + \left(\frac{2h}{P}\right)^2\left(z^2 - \left(\frac{P}{2h}\right)^2\right) = \left(\frac{2h}{P}\right)^2 z^2.$$

Using this result and equation (B.5), we get

$$z = \frac{P}{2h}\cosh H = r + \frac{\mu}{2h},$$

which results in

$$r = -\frac{\mu}{2h} + \frac{P}{2h}\cosh H$$

$$= -\frac{\mu}{2h}\left(1 - \frac{P}{\mu}\cosh H\right).$$
(B.10)

Compare equation (B.10) with equation (2.37).

We recall here some characteristics of hyperbolic functions (see also fig. B.1):

$$\sinh x = \frac{1}{2}\left(e^x - e^{-x}\right), \quad -\infty < \sinh x < \infty, \ \sinh(0) = 0$$

$$\cosh x = \frac{1}{2}\left(e^x + e^{-x}\right), \quad 1 \le \cosh x < \infty, \ \cosh(0) = 1$$

$$d(\sinh x) = \cosh x\,dx$$

$$d(\cosh x) = \sinh x\,dx.$$

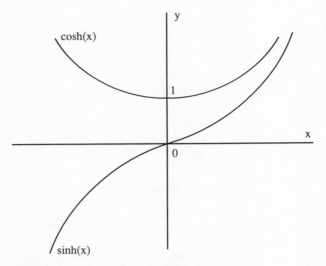

Figure B.1   Graphs of $\sinh(x)$ and $\cosh(x)$.

Equation (B.10) becomes

$$r = a(1 - e \cosh H).$$ (B.11)

Compare this equation with equation (4.5), where we note that since $a < 0$, in order for $r > 0$, we must have $e \geq 1$.

Equation (B.9) becomes

$$\sqrt{\frac{\mu}{-a^3}}\,(t + K) = e \sinh H - H.$$ (B.12)

Comparing this equation with equation (4.1), we can apply the condition $t = t_\pi$ when $H = 0$. Therefore,

$$K = -t_\pi.$$ (B.13)

Note that if we take the time derivative of equation (B.12),

$$\sqrt{\frac{\mu}{-a^3}}\,\frac{dt}{dH} = e \cosh H - 1,$$

which from equation (B.11) yields

$$\sqrt{\frac{\mu}{-a^3}}\,\frac{dt}{dH} = -\frac{r}{a}$$

or

$$\frac{dH}{dt} = -\frac{a}{r}\sqrt{\frac{\mu}{-a^3}} = -\frac{1}{r}\sqrt{\frac{\mu}{-a}}. \qquad (B.14)$$

Compare equation (B.14) with equation (4.12).

Note from §4.1 that the equation of conic motion remains the same. That is,

$$r = \frac{p}{1 + e\cos\phi},$$

where

$$p = \frac{c^2}{\mu} = a(1 - e^2)$$

$$e = \frac{P}{\mu}$$

$$a = -\frac{\mu}{2h}.$$

C

CONIC SECTIONS

The general equation of a conic has the form

$$r = \frac{p}{1 + e\cos\phi},$$

where

$p \equiv$ semi-latus rectum > 0

$e \equiv$ eccentricity

$r, \phi \equiv$ the polar coordinates of a point on the conic, and

$\phi \equiv$ true anomaly in celestial and orbital mechanics.

For the case of the ellipse which is shown in figure C.1,

F is the Focus

F' is the Vacant (or Empty) Focus

$p = a(1 - e^2), \ a > 0$

$AF + AF' = 2a$

$e = \sqrt{1 - (b/a)^2} \leq 1$

a is the semi-major axis

b is the semi-minor axis.

For the case of the hyperbola, which is shown in figure C.2,

$p = a(1 - e^2), \ a < 0$

$AF - AF' = 2a$

$e \geq 1.$

Note that as $r \to \infty$, the curves approach the asymptotes, as shown in figure C.3. And finally, for the case of the parabola that is shown in figure C.4,

$q = r_\pi$

$p = 2q, \ a \to \infty$

$AN = AF$ at *all* points along the parabola

$e = 1.$

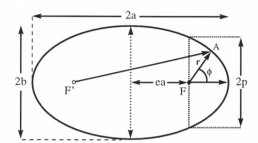

Figure C.1   Conic section parameters for the ellipse.

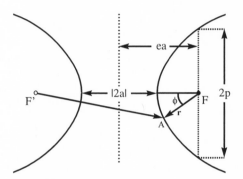

Figure C.2   Conic section parameters for the hyperbola.

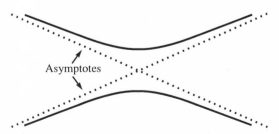

Figure C.3   Hyperbolic curves approaching asymptotes.

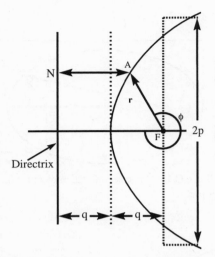

Figure C.4   Conic section parameters for the parabola.

D

TRANSFER-ANGLE RESOLUTION

In this appendix we resolve the ambiguity in transfer angle $\Delta\phi$ in Lambert's problem and LTVCON. Let $\mathbf{r}_0$ and $\mathbf{r}$ be specified in the inertial Cartesian frame x, y, z (see fig. D.1).

From $\mathbf{r}_0$ and $\mathbf{r}$ we can compute the quantity

$$q = \hat{\mathbf{k}} \cdot \mathbf{r}_0 \times \mathbf{r},$$

which is the z-component of $\mathbf{r}_0 \times \mathbf{r}$. The quantity q can be expanded in terms of the transfer angle ($\Delta\phi$) and the inclination

$$q = \hat{\mathbf{k}} \cdot \hat{\mathbf{c}} \, r_0 r \sin \Delta\phi$$
$$= r_0 \, r \cos i \sin \Delta\phi.$$

If $q > 0$, then *either*

$$0 < i < \frac{\pi}{2} \quad \text{and} \quad 0 < \Delta\phi < \pi,$$

which results in

$$\cos i > 0 \quad \text{and} \quad \sin \Delta\phi > 0,$$

or

$$\frac{\pi}{2} < i < \pi \quad \text{and} \quad \pi < \Delta\phi < 2\pi,$$

which results in

$$\cos i < 0 \quad \text{and} \quad \sin \Delta\phi < 0.$$

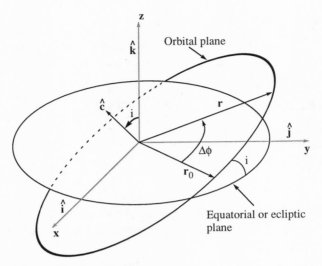

Figure D.1 Inertial system x, y, z.

Therefore, for a *direct* orbit ($0 < i < \pi/2$),

$$\Delta\phi = \cos^{-1}\left(\frac{\mathbf{r}_0 \cdot \mathbf{r}}{r_0 r}\right)$$

$$\cos i > 0$$

$$\sin \Delta\phi > 0$$

yields $0 < \Delta\phi < \pi$. For a *retrograde* orbit ($\pi/2 < i < \pi$),

$$\Delta\phi = 2\pi - \cos^{-1}\left(\frac{\mathbf{r}_0 \cdot \mathbf{r}}{r_0 r}\right)$$

$$\cos i < 0$$

$$\sin \Delta\phi < 0$$

yields $\pi < \Delta\phi < 2\pi$. However, if $q < 0$, then *either*

$$0 < i < \frac{\pi}{2} \quad \text{and} \quad \pi < \Delta\phi < 2\pi,$$

which results in

$$\cos i > 0 \quad \text{and} \quad \sin \Delta\phi < 0,$$

or

$$\frac{\pi}{2} < i < \pi \quad \text{and} \quad 0 < \Delta\phi < \pi,$$

which results in

$$\cos i < 0 \quad \text{and} \quad \sin \Delta\phi > 0.$$

Therefore, for a *direct* orbit $(0 < i < \pi/2)$,

$$\Delta\phi = 2\pi - \cos^{-1}\left(\frac{\mathbf{r}_0 \cdot \mathbf{r}}{r_0 r}\right)$$

$$\cos i > 0$$

$$\sin \Delta\phi < 0$$

yields $\pi < \Delta\phi < 2\pi$. For a *retrograde* orbit $(\pi/2 < i < \pi)$,

$$\Delta\phi = \cos^{-1}\left(\frac{\mathbf{r}_0 \cdot \mathbf{r}}{r_0 r}\right)$$

$$\cos i < 0$$

$$\sin \Delta\phi > 0$$

yields $0 < \Delta\phi < \pi$. Thus the ambiguity is resolved by choosing a *direct* or *retrograde* orbit *a priori* and checking the sign of

$$q = (\mathbf{r}_0 \times \mathbf{r})_z.$$

E

STUMPFF FUNCTIONS

The Stumpff functions [66] and [64] are defined by the series

$$c_n(z) = \sum_{k=0}^{\infty} (-1)^k \frac{z^k}{(2k+n)!},$$

where $n = 0, 1, 2, 3, \ldots$. Let $z = \alpha_J s^2$; then the first few of these functions are

$$c_0 = c_0(z) = \begin{cases} \cos \sqrt{z} & \alpha_J > 0 \\ \cosh \sqrt{-z} & \alpha_J < 0 \\ 1 & \alpha_J = 0 \end{cases}$$

$$c_1 = c_1(z) = \begin{cases} \frac{\sin \sqrt{z}}{\sqrt{z}} & \alpha_J > 0 \\ \frac{\sinh \sqrt{-z}}{\sqrt{-z}} & \alpha_J < 0 \\ 1 & \alpha_J = 0 \end{cases}$$

$$c_2 = c_2(z) = \begin{cases} \frac{(1-\cos \sqrt{z})}{z} & \alpha_J > 0 \\ \frac{(\cosh \sqrt{-z}-1)}{-z} & \alpha_J < 0 \\ \frac{1}{2} & \alpha_J = 0 \end{cases}.$$

We see that this *single* class of infinite series represents both *trigonometric* ($\alpha_J > 0$) and *hyperbolic* ($\alpha_J < 0$) functions. The *parabolic* ($\alpha_J = 0$) case is also included, since

$$c_n(0) = \frac{1}{n!}.$$

Observe that from the definition,

$$c_n(z) + z c_{n+2}(z) = \frac{1}{n!}. \qquad (E.1)$$

The Stumpff functions have convenient derivative formulae. For example,

$$2z\frac{dc_n(z)}{dz} = c_{n-1}(z) - nc_n(z), n > 0$$

and

$$\frac{dc_n(z)}{dz} = \frac{1}{2}[nc_{n+2}(z) - c_{n+1}(z)], \quad n \geq 0.$$

For the case where $z = \alpha$, s^2, we have

$$\frac{dz}{ds} = 2\alpha, s$$

and the *first* derivative formula becomes

$$2\alpha, s^2\frac{dc_n(\alpha, s^2)}{dz}\frac{dz}{ds} = \left[c_{n-1}(\alpha, s^2) - nc_n(\alpha, s^2)\right]2\alpha, s$$

or

$$s\frac{dc_n(\alpha, s^2)}{ds} = c_{n-1}(\alpha, s^2) - nc_n(\alpha, s^2). \tag{E.2}$$

The *second* derivative formula becomes

$$\frac{dc_n(\alpha, s^2)}{dz}\frac{dz}{ds} = \frac{1}{2}\left[nc_{n+2}(\alpha, s^2) - c_{n+1}(\alpha, s^2)\right]2\alpha, s$$

or

$$\frac{dc_n(\alpha, s^2)}{ds} = \alpha, s\left[nc_{n+2}(\alpha, s^2) - c_{n+1}(\alpha, s^2)\right]. \tag{E.3}$$

If we understand that the argument to the Stumpff functions is α, s^2 and $()' = \frac{d}{ds}$, these two derivative equations become

$$sc_n' = c_{n-1} - nc_n, \quad n > 0 \tag{E.4}$$

$$c_n' = \alpha, s\,(nc_{n+2} - c_{n+1}), \quad n \geq 0. \tag{E.5}$$

We also have an identity for integration,

$$\int s^k c_k(\rho s^2)ds = s^{k+1}c_{k+1}(\rho s^2), \tag{E.6}$$

where ρ is arbitrary. The first few of these are as follows:

$$k = 0: \quad \int c_0 ds = sc_1 \tag{E.7}$$

$$k = 1: \quad \int sc_1 ds = s^2c_2 \tag{E.8}$$

$$k = 2: \int s^2c_2 ds = s^3c_3. \tag{E.9}$$

Some other important identities are

$$c_0^2(z) + zc_1^2(z) = 1 \tag{E.10}$$

$$c_0^2(z) - zc_1^2(z) = c_0(4z) \tag{E.11}$$

$$c_0^2(z) = 1 - 2zc_2(4z) \tag{E.12}$$

$$c_1^2(z) = 2c_2(4z) \tag{E.13}$$

$$c_1(4z) = c_0(z)\,c_1(z). \tag{E.14}$$

F

ORBIT GEOMETRY

In this appendix we introduce a geometrical interpretation of the unit vectors along the angular momentum and the line of apsides. Define the inertial system x, y, and z at the fixed origin (one of the masses; see fig. F.1).

The parameters illustrated in figure F.1 are as follows:

Ω	Ascending node ($0 \leq \Omega \leq 2\pi$)
i	Orbit inclination ($0 \leq i \leq \pi$)
$\hat{\mathbf{P}}$	Laplace unit vector
$\hat{\mathbf{c}}$	Angular momentum unit vector
ω	Argument of pericenter
π	Periapsis point
$\hat{\mathbf{i}}, \hat{\mathbf{j}}, \hat{\mathbf{k}}$	Orthogonal unit vectors along x, y, z.

We now develop expressions for $\hat{\mathbf{c}}$ and $\hat{\mathbf{P}}$ as functions of Ω, i, and ω. From figure F.1 we see that the unit vector along the nodal line is defined by

$$\hat{\mathbf{k}} \times \hat{\mathbf{c}} = \hat{\mathbf{N}} \sin i. \tag{F.1}$$

In the equatorial plane, this unit vector is given by

$$\hat{\mathbf{N}} = \hat{\mathbf{i}} \cos \Omega + \hat{\mathbf{j}} \sin \Omega. \tag{F.2}$$

We can determine the components of the unit vector $\hat{\mathbf{c}}$ as follows. From the vector triple product,

$$\hat{\mathbf{i}} \times (\hat{\mathbf{k}} \times \hat{\mathbf{c}}) = (\hat{\mathbf{i}} \cdot \hat{\mathbf{c}})\hat{\mathbf{k}} - (\hat{\mathbf{i}} \cdot \hat{\mathbf{k}})\hat{\mathbf{c}} = (\hat{\mathbf{i}} \cdot \hat{\mathbf{c}})\hat{\mathbf{k}}.$$

From equations (F.1) and (F.2), we get

$$(\hat{\mathbf{i}} \cdot \hat{\mathbf{c}})\hat{\mathbf{k}} = \hat{\mathbf{i}} \times \hat{\mathbf{N}} \sin i,$$

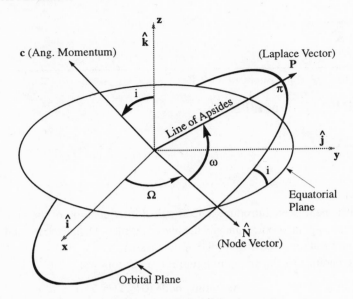

Figure F.1   Orbital elements in the inertial system x, y, z.

so

$$(\hat{\mathbf{i}} \cdot \hat{\mathbf{c}})\hat{\mathbf{k}} = \hat{\mathbf{i}} \times (\hat{\mathbf{i}} \cos \Omega + \hat{\mathbf{j}} \sin \Omega) \sin i.$$

Therefore,

$$(\hat{\mathbf{i}} \cdot \hat{\mathbf{c}}) = \sin \Omega \sin i. \tag{F.3}$$

Similarly we can get $\hat{\mathbf{j}} \cdot \hat{\mathbf{c}}$. From the vector triple product,

$$\hat{\mathbf{j}} \times (\hat{\mathbf{k}} \times \hat{\mathbf{c}}) = (\hat{\mathbf{j}} \cdot \hat{\mathbf{c}})\hat{\mathbf{k}} - (\hat{\mathbf{j}} \cdot \hat{\mathbf{k}})\hat{\mathbf{c}} = (\hat{\mathbf{j}} \cdot \hat{\mathbf{c}})\hat{\mathbf{k}}.$$

So

$$(\hat{\mathbf{j}} \cdot \hat{\mathbf{c}})\hat{\mathbf{k}} = \hat{\mathbf{j}} \times \hat{\mathbf{N}} \sin i.$$

Therefore,

$$\hat{\mathbf{j}} \cdot \hat{\mathbf{c}} = -\cos \Omega \sin i, \tag{F.4}$$

and from figure F.1,

$$\hat{\mathbf{k}} \cdot \hat{\mathbf{c}} = \cos i. \tag{F.5}$$

Collecting equations (F.3)–(F.5) we construct the unit angular momentum,

$$\hat{\mathbf{c}} = \hat{\mathbf{i}} \sin \Omega \sin i - \hat{\mathbf{j}} \cos \Omega \sin i + \hat{\mathbf{k}} \cos i. \qquad (\text{F.6})$$

To find the components of the unit vector $\hat{\mathbf{P}}$, we take the cross product of the node vector $(\hat{\mathbf{N}})$ and $\hat{\mathbf{P}}$. From figure F.1,

$$\hat{\mathbf{N}} \times \hat{\mathbf{P}} = \sin \omega \, \hat{\mathbf{c}}. \qquad (\text{F.7})$$

We now dot this with $\hat{\mathbf{i}}$ and use equation (F.3) to obtain

$$\hat{\mathbf{i}} \times \hat{\mathbf{N}} \cdot \hat{\mathbf{P}} = \sin \omega \hat{\mathbf{i}} \cdot \hat{\mathbf{c}} = \sin \omega \sin \Omega \sin i.$$

Substituting for $\hat{\mathbf{N}}$ using equation (F.2), we obtain

$$(\hat{\mathbf{i}} \times \hat{\mathbf{j}} \sin \Omega) \cdot \hat{\mathbf{P}} = \sin \omega \sin \Omega \sin i.$$

Cancel the $\sin \Omega$ on both sides and use $\hat{\mathbf{i}} \times \hat{\mathbf{j}} = \hat{\mathbf{k}}$ to arrive at

$$\hat{\mathbf{k}} \cdot \hat{\mathbf{P}} = \sin \omega \sin i. \qquad (\text{F.8})$$

In §2.4.4 we developed the relation

$$\hat{\mathbf{c}} \cdot \hat{\mathbf{P}} = 0,$$

which becomes, using equation (F.6),

$$\hat{\mathbf{i}} \cdot \hat{\mathbf{P}} \sin \Omega \sin i - \hat{\mathbf{j}} \cdot \hat{\mathbf{P}} \cos \Omega \sin i + \hat{\mathbf{k}} \cdot \hat{\mathbf{P}} \cos i = 0.$$

From equation (F.8) the $\sin i$ will cancel to give

$$(\hat{\mathbf{i}} \cdot \hat{\mathbf{P}}) \sin \Omega - (\hat{\mathbf{j}} \cdot \hat{\mathbf{P}}) \cos \Omega + \sin \omega \cos i = 0. \qquad (\text{F.9})$$

To obtain a *second* relation involving $\hat{\mathbf{i}} \cdot \hat{\mathbf{P}}$ and $\hat{\mathbf{j}} \cdot \hat{\mathbf{P}}$, we form the cross product

$$\hat{\mathbf{c}} \times \hat{\mathbf{P}} = \hat{\mathbf{c}} \times \left[\hat{\mathbf{i}}(\hat{\mathbf{i}} \cdot \hat{\mathbf{P}}) + \hat{\mathbf{j}}(\hat{\mathbf{j}} \cdot \hat{\mathbf{P}}) + \hat{\mathbf{k}}(\hat{\mathbf{k}} \cdot \hat{\mathbf{P}}) \right]. \qquad (\text{F.10})$$

Dot this equation with $\hat{\mathbf{k}}$. Since $\hat{\mathbf{k}} \cdot \hat{\mathbf{c}} \times \hat{\mathbf{k}} = 0$ we obtain

$$\hat{\mathbf{k}} \cdot \hat{\mathbf{c}} \times \hat{\mathbf{P}} = \hat{\mathbf{k}} \cdot \hat{\mathbf{c}} \times \hat{\mathbf{i}}(\hat{\mathbf{i}} \cdot \hat{\mathbf{P}}) + \hat{\mathbf{k}} \cdot \hat{\mathbf{c}} \times \hat{\mathbf{j}}(\hat{\mathbf{j}} \cdot \hat{\mathbf{P}}).$$

Now interchange the dot and cross products in the scalar triple products on the left side,

$$\hat{\mathbf{k}} \times \hat{\mathbf{c}} \cdot \hat{\mathbf{P}} = \hat{\mathbf{c}} \times \hat{\mathbf{i}} \cdot \hat{\mathbf{k}}(\hat{\mathbf{i}} \cdot \hat{\mathbf{P}}) + \hat{\mathbf{c}} \times \hat{\mathbf{j}} \cdot \hat{\mathbf{k}}(\hat{\mathbf{j}} \cdot \hat{\mathbf{P}}),$$

and also use equation (F.1) on the left side,

$$\hat{\mathbf{N}} \cdot \hat{\mathbf{P}} \sin i = \hat{\mathbf{c}} \times \hat{\mathbf{i}} \cdot \hat{\mathbf{k}}(\hat{\mathbf{i}} \cdot \hat{\mathbf{P}}) + \hat{\mathbf{c}} \times \hat{\mathbf{j}} \cdot \hat{\mathbf{k}}(\hat{\mathbf{j}} \cdot \hat{\mathbf{P}}).$$

Observe that $\hat{\mathbf{N}} \cdot \hat{\mathbf{P}} = \cos \omega, \hat{\mathbf{i}} \times \hat{\mathbf{k}} = -\hat{\mathbf{j}}$ and $\hat{\mathbf{j}} \times \hat{\mathbf{k}} = \hat{\mathbf{i}}.$ Using these relationships in the above equation results in

$$\cos \omega \sin i = -\hat{\mathbf{j}} \cdot \hat{\mathbf{c}}(\hat{\mathbf{i}} \cdot \hat{\mathbf{P}}) + \hat{\mathbf{i}} \cdot \hat{\mathbf{c}}(\hat{\mathbf{j}} \cdot \hat{\mathbf{P}}).$$

We now use equation (F.3) and (F.4) and reverse sides to get

$$\cos \Omega \sin i (\hat{\mathbf{i}} \cdot \hat{\mathbf{P}}) + \sin \Omega \sin i (\hat{\mathbf{j}} \cdot \hat{\mathbf{P}}) = \cos \omega \sin i.$$

Cancel the $\sin i$ term to obtain

$$(\hat{\mathbf{i}} \cdot \hat{\mathbf{P}}) \cos \Omega + (\hat{\mathbf{j}} \cdot \hat{\mathbf{P}}) \sin \Omega = \cos \omega. \tag{F.11}$$

Now we solve equations (F.11) and (F.9) simultaneously for $\hat{\mathbf{i}} \cdot \hat{\mathbf{P}}$ and $\hat{\mathbf{j}} \cdot \hat{\mathbf{P}}.$ Multiply equation (F.11) by $\cos \Omega$ and equation (F.9) by $\sin \Omega$ and add to obtain

$$\hat{\mathbf{i}} \cdot \hat{\mathbf{P}} = \cos \omega \cos \Omega - \sin \omega \sin \Omega \cos i. \tag{F.12}$$

Eliminate $(\hat{\mathbf{i}} \cdot \hat{\mathbf{P}})$ from equation (F.11) by substitution of equation (F.12) to obtain

$$\hat{\mathbf{j}} \cdot \hat{\mathbf{P}} = \cos \omega \sin \Omega + \sin \omega \cos \Omega \cos i. \tag{F.13}$$

Equations (F.8), (F.12), and (F.13) yield the unit vector

$$\hat{\mathbf{P}} = \hat{\mathbf{i}}(\cos \omega \cos \Omega - \sin \omega \sin \Omega \cos i)$$
$$+ \hat{\mathbf{j}}(\cos \omega \sin \Omega + \sin \omega \cos \Omega \cos i) + \hat{\mathbf{k}}(\sin \omega \sin i). \tag{F.14}$$

REFERENCES

[1] Arfken, G. *Mathematical Methods for Physicists*. Academic Press, New York, 1985.

[2] The Astronomical Almanac for Year 1985. Prepared by the U.S. Naval Observatory, Washington, D.C., and the Royal Greenwich Observatory, East Sussex, England, 1985.

[3] Bate, R., Mueller, D., and White, J. *Fundamentals of Astrodynamics*. Dover, New York, 1971.

[4] Battin, R. H. *Astronautical Guidance*. McGraw-Hill, New York, 1964.

[5] Battin, R. H. *An Introduction to the Mathematics and Methods of Astrodynamics*. AIAA Education Series, 1987.

[6] Bettis, D. G. Efficient Embedded Runge-Kutta Methods. TICOM Report 77-12, the Texas Institute for Computational Mechanics, the University of Texas at Austin, November 1977.

[7] Blitzer, L., et al. Effect of Ellipticity on 24-Hour Nearly Circular Satellite Orbits. *Journal of Geophysical Research*, 67(1), January 1963.

[8] Bogdan, V., and Bond, V. R. Global Estimate of Deviation of a Solution of a System of Differential Equations Due to a Perturbation. In *Final Program, Society for Industrial and Applied Mathematics (SIAM)*, Alexandria, Virginia, National Meeting, 1980.

[9] Bond, V., and Anson, K. W. Trajectories That Flyby Jupiter and Saturn and Return to Earth. *Journal of Spacecraft and Rockets*, 9, 1972.

[10] Bond, V., Fraietta, M., and Sponaugle, S. A Modern Application of the Jacobian Integral Using a Special Perturbation Method. In Edward Belbruno, ed., *Proceedings on Advances in Nonlinear Astrodynamics*, Preprint Series No. GCG65, presented at the Geometry Center of the National Science Foundation at the University of Minnesota, Minneapolis, November 8–10, 1993.

[11] Bond, V. R. A Two-Body Analysis of ΔV Requirements Necessary for the Abort from a Translunar Mission. MSC Internal Note 64-EG-3, NASA/Manned Spacecraft Center Publication, March 1964.

[12] Bond, V. R. Recursive Computation of the Coefficients of the Time-Dependent f and g Series Solution of Keplerian Motion and a Study of the Convergence Properties of the Solution. MSC Internal Note TN D-3322, NASA/Manned Spacecraft Center Publication, March 1966.

[13] Bond, V. R. A Recursive Formulation for Computing the Coefficients of the Time-Dependent f and g Series Solution to the Two-Body Problem. *Astronomical Journal*, 71(1), 1966.

[14] Bond, V. R. The Uniform, Regular Differential Equations of the KS Transformed Perturbed Two-Body Problem. *Celestial Mechanics*, 10, 1974.

[15] Bond, V. R. Error Propagation in the Numerical Solutions of the Differential Equations of Orbital Mechanics. *Celestial Mechanics*, 27, 1982.

[16] Bond, V. R. The Perturbative Effects of a Tethered Satellite on the Orbital Elements. In *11th Annual AIAA Houston Section Technical Symposium*, Houston, Texas, May 1986.

[17] Bond, V. R., and Fraietta, M. F. Elimination of Secular Terms from the Differential Equations for the Elements of the Perturbed Two-Body Problem. In *Proceedings of the Flight Mechanics and Estimation Theory Symposium—1991 (NASA CP 3123)*, Greenbelt, Maryland, May 1991. NASA/Goddard Space Flight Center.

[18] Bond, V. R., and Gottlieb, R. G. A Perturbed Two-Body Element Method Utilizing the Jacobian Integral. JSC Internal 73-FM-86 (JSC-08004), NASA/Johnson Space Center (JSC), June 1989.

[19] Bond, V. R., and Hanssen, V. The Burdet Formulation of the Perturbed Two-Body Problem with Total Energy as an Element. JSC Internal 73-FM-86 (JSC-08004), NASA/Johnson Space Center (JSC), June 1973.

[20] Bond, V. R., and Henry, E. W. One-Way and Flyby Interplanetary Trajectory Approximations Using Matched Conic Techniques. MSC Internal 67-FM-1667, NASA/Manned Spacecraft Center, 1967.

[21] Bond, V. R., and Horn, M. K. Error Propagation in Orbital Motion Equations Using Runge-Kutta Methods. JSC Internal TM 58216, NASA/Johnson Space Center (JSC), March 1979.

[22] Bond, V. R., and Mulcihy, D. D. Computation of Orbits Using Total Energy. In *Proceedings of the Flight Mechanics and Estimation Theory Symposium—1988 (NASA CP 3011)*, Greenbelt, Maryland, May 1988. NASA/Goddard Space Flight Center.

[23] Bronshtein, I. N., and Semendyayev, K. A. *Handbook of Mathematics*. Van Nostrand Reinhold, New York, 1985.

[24] Brouwer, D., and Clemence, G. M. *Methods of Celestial Mechanics*. Academic Press, New York, 1961.

[25] Burdet, C. A. Theory of Kepler Motion. *Zeitschrift für Angewandte Mathematik und Physik*, 19, 1968.

[26] Campbell, J., Moore, W., and Wolf, H. A General Method for Selection and Optimization of Trajectories. In R. L. Duncombe and V. G. Szebehely, eds., *Methods in Astrodynamics and Celestial Mechanics*. Academic Press, New York, 1966.

[27] Constant, F. W. *Theoretical Physics*. Addison-Wesley, Reading, Mass., 1954.

[28] Danby, J. M. A. *Fundamentals of Celestial Mechanics*. Macmillan, New York, 1962.

[29] Danby, J. M. A. Transformations to Extend the Range of the Application of Power Series Solutions of the Differential Equations of Motion. *Celestial Mechanics*, 5, 1972.

[30] Davis, H. F., and Snider, A. D. *Introduction to Vector Analysis*. Wm. C. Brown, Dubugue, Iowa, 1988.

[31] The Explanatory Supplement to the Ephemeris. Prepared by the Nautical Almanac Offices of the United Kingdom and the United States of America, 1961.

[32] Fehlberg, E. Low Order Classical Runge-Kutta Formulas with Stepsize Control and Their Application to Some Heat Transfer Problems. Technical Report, NASA Technical Note 315, 1969. Also in *Computing*, 6, 1970, in German.

[33] Giacaglia, G. E. O. *Perturbation Methods in Non-Linear Systems*. Springer-Verlag, New York, 1972.

[34] Goldstein, H. *Classical Mechanics*. Addison-Wesley, Reading, Mass., 1981.

[35] Gottlieb, R. G. Personal communication to the author (VRB), February 1991.

[36] Gottlieb, R. G. Fast Gravity, Gravity Partials, Normalized Gravity, Gravity Gradient Torque and Magnetic Field: Derivation, Code and Data. JSC Internal Note CR-188243, NASA/Johnson Space Center (JSC), January 1993.

[37] Graf, O. The Elimination of Short and Intermediate Period Terms from the Problem of a High Altitude Earth Satellite. *Celestial Mechanics*, 14, 1976.

[38] Green, R. M. *Spherical Astronomy*. Cambridge University Press, Cambridge, England, 1985.

[39] Horn, M. K. Scaled Runge-Kutta Algorithms for Treating the Problem of Dense Output. JSC Internal TM 58239, NASA/Johnson Space Center (JSC), March 1982.

[40] International Earth Rotation Service Bulletin A. Distributed Weekly by the U.S. Naval Observatory, Washington, D.C.

[41] The International System of Units (SI). NBS Special Publication 330. U.S. Dept. of Commerce, National Bureau of Standards, 1986.

[42] Janin, G., and Bond, V. R. A General Time Element for Orbit Integration in Cartesian Coordinates. *Advances in Space Research*, 1, COSPAR, 1981.

[43] Johnson, G. The Unique Determination of Density from Higher-Order Potentials. JSC Internal 80-FM-42 (JSC-16771), NASA/Johnson Space Center (JSC), 1980.

[44] Johnson, L. W., and Reiss, R. D. *Introduction to Linear Algebra*. Addison-Wesley, Reading, Mass., 1981.

[45] Kaplan, G. H. The IAU Resolutions on the Astronomical Constants, Time Scales and the Fundamental Reference Frame. Circular 163, United States Naval Observatory, Washington, D.C., December 1981.

[46] Kennedy, E. Lyapunov Stability and Its Application to Systems of Ordinary Differential Equations. JSC Internal Note 79-FM-31 (JSC-16079), NASA/Johnson Space Center (JSC), September 1979.

[47] Kreyszig, E. *Advanced Engineering Mathematics*. John Wiley and Sons, New York, 1988.

[48] Lanczos, C. *The Variational Principles of Mechanics*. Dover, New York, 1986.

[49] Landau, L. D., and Lifshitz, E. M. *Mechanics: Course in Theoretical Physics*, vol. 1. Pergamon Press, Elmsford, N.Y., 1989.

[50] Lindsay, H., and Margenau, R. *Foundations of Physics*. Dover, New York, 1957.

[51] Long, A. D. Personal communication to author (VRB), August 1991. NASA/Johnson Space Center.

[52] Marsden, J. E., and Tromba, A. J. *Vector Calculus*. W. H. Freeman, New York, 1981.

[53] Meirovitch, L. *Methods of Analytical Dynamics*. McGraw-Hill, New York, 1970.

[54] Moulton, F. R. *An Introduction to Celestial Mechanics*. Macmillan, New York, 1914. (This reference republished by Dover, New York, 1970.)

[55] Mueller, A. C. A Fast Recursive Algorithm for Calculating the Forces Due to the Geopotential (Program: GEOPOT). JSC Internal Note 75-FM-42 (JSC-09731), NASA/Johnson Space Center (JSC), 1975.

[56] Murad, P. A. Tsien's Method for Generating Non-Keplerian Trajectories. Presented at the 29th Aerospace Sciences Meeting, Reno, Nevada, January 1991. AIAA.

[57] Nacozy, P. The Lyapunov Stabilization of Satellite Equations of Motion Using Integrals. In *Proceedings of the AIAA/AAS Astrodynamics Conference*, July 1973, Vail, Colorado, 1973.

[58] Pollard, H. *Mathematical Introduction to Celestial Mechanics*. Prentice-Hall, 1966.

[59] Seidelmann, P. K., Ed. *Explanatory Supplement to the Astronomical Almanac*. University Science Books, Mill Valley, Calif., 1991.

[60] Shampine, L. F., and Gordon, M. K. *Computer Solution of Ordinary Differential Equations: The Initial Value Problem*. W. H. Freeman, San Francisco, 1975.

[61] Sperling, H. Computation of Keplerian Conic Sections. *ARS Journal*, May 1961.

[62] Stakgold, I. *Green's Functions and Boundary Value Problems*. John Wiley and Sons, New York, 1979.

[63] Stiefel, E. L., and Baumgarte, J. Examples of Transformations Improving the Numerical Accuracy of the Integration of Differential Equations. In *Proceedings of the Conference on Numerical Solutions of Ordinary Differential Equations, October, 1972*, Austin, Texas, 1972.

[64] Stiefel, E. L., and Scheifele, G. *Linear and Regular Celestial Mechanics*. Springer-Verlag, New York, 1971.

[65] Stoer, J., and Bulirsch, R. *Introduction to Numerical Analysis*. Texts in Applied Mathematics 12, 2d ed. Springer-Verlag, New York, 1993.

[66] Stumpff, K. Neue Formeln und Hilfstafeln zur Ephemeridenrechnung. *Astronomische Nachrichten*, 275, 1947.

[67] Sundmann, K. Memoire sur le problème des trois corps. *Acta Mathematika*, 36, 1912.

[68] Symon, K. R. *Mechanics*. Addison-Wesley, Reading, Mass., 1971.

[69] Szebehely, V. *Theory of Orbits: The Restricted Problem of Three Bodies*. Academic Press, New York-London, 1967.

[70] Szebehely, V. *Adventures in Celestial Mechanics*. University of Texas Press, Austin, 1989.

[71] Szebehely, V., and Bond, V. R. Transformations of the Perturbed Two-Body Problem to Unperturbed Oscillators. *Celestial Mechanics*, 30, 1983.

[72] Taff, L. G. *Celestial Mechanics: A Computational Guide for the Practitioner*. John Wiley and Sons, New York, 1985.

[73] Thomas, B. The Runge-Kutta Methods. *BYTE*, April 1986.

[74] U.S. Naval Observatory, Washington, D.C. Time Service Announcement Series 14, No. 56, January 25, 1994.

[75] Van Flandern, T., and Pulkkinen, K. Low-Precision Formulae for Planetary Positions. *Astrophysical Journal Supplement Series*, 41, November 1979. (Also reprinted by William Bell, Inc., Richmond, Va.)

[76] Volk, O. Kepleriana. *Celestial Mechanics*, 8, 1973.

[77] Wolfram, S. *Mathematica: A System for Doing Mathematics by Computer*. 2d ed. Addison-Wesley, Reading, Mass., 1991.

INDEX